AF885316

ELEVATE

FOR VETERANS

Mindset, Emotional, and Financial Mastery for Life After Service

DR. RODERICK C. CUNNINGHAM, CMSgt (Ret.)

VALERIE N. CUNNINGHAM

Published 2025 by Chief Empowerment Network, LLC.

For customized training materials and/or to become a certified trainer of this material, contact Chief Empowerment Network, LLC directly at 813-380-8045.

Contributor: Dexter T. Wyckoff, Financial Advisor

Book Cover Layout: Valerie Cunningham and Xie Maxine Guiliane

Inside Photography: Katwon Turner Photography

Edited by: Valerie N. Cunningham and Dr. Roderick C. Cunningham

Published by: **Chief Empowerment Network LLC | www.FromPages2Stages.com**

Printed in the United States of America

ISBN 978-1-7322659-8-1

For additional resources, go to

www.BookRod.com

IG: @RodTheSpeaker

LinkedIn: www.Linkedin.com/roderickcc

Table of Contents

Dedication

This book is lovingly dedicated to all the men and women who have served in the Armed Forces. Your sacrifice, courage, and dedication to our country have inspired us to create this resource specifically designed to support your transition and continued success in civilian life.

"You stood the watch so others could sleep peacefully."

To the families who waited, worried, and welcomed home our warriors, your strength and resilience are the foundation upon which our service members stand.

And to those who made the ultimate sacrifice, we honor your memory by supporting those who returned and continue to carry your spirit with them.

With profound respect and gratitude,

Rod and Valerie

Acknowledgments

First and foremost, we extend our deepest gratitude to the veterans who shared their transition stories with us. Your candid insights about the challenges and triumphs of moving from military to civilian life have been instrumental in shaping this workbook. Your willingness to share both struggles and successes ensures that this resource addresses real needs with practical solutions.

To our military and veteran service organizations, VA, American Legion, Disabled American Veterans, Veterans of Foreign Wars, Wounded Warrior Project, and countless others, your tireless work supporting our nation's heroes has informed our approach and deepened our understanding of the veteran experience.

We are profoundly thankful to our academic partners who have committed to veteran success through education and employment opportunities (like St. Petersburg College, Hillsborough College, and Miami Dade College). Your dedication to recognizing and developing the unique talents that veterans bring to the workforce has created pathways for countless service members to thrive in civilian careers.

Our journey is also made possible by the innumerable mentors, colleagues, and friends who have guided and walked alongside us. Your wisdom and encouragement are the scaffolding upon which we build our dreams and ambitions.

To our beloved children, Dexter, Erika, and Amber, and our cherished eight grandchildren (Jordyn, Jeremiah, Trenton, Zakai, Jayson, Miya, Amelia, and Sebastian), your endless support and love illuminate our path. You inspire us to set examples of strength, resilience, and compassion through our work. We hope that our efforts will be reflected in our legacy for you, fostering a world enriched with kindness and understanding.

Lastly, to all those who share our vision and support our work, whether by reading our books, attending our seminars, or spreading our message, you are integral to our success. Together, we forge a path toward a more understanding, empathetic, and connected world for our veterans.

With heartfelt thanks,

Rod and Valerie

About the Authors

Dr. Roderick C. Cunningham and **Valerie Cunningham** are life partners and a dynamic duo in the world of motivational speaking, authorship, and emotional and financial wellness. With over 30 years of marriage, three children, and eight grandchildren, their journey together has been rich in love, leadership, and service.

Dr. Roderick C. Cunningham

A retired United States Air Force Chief Master Sergeant (CMSGT) with 29 years of distinguished service (20 years active duty and 9 years reserve), Dr. Rod Cunningham has dedicated his life to leadership and emotional wellness. His military career began with 11.5 years of active duty, serving at Ramstein AB, Germany; Eaker AFB, Arkansas; Shemya AFB, Alaska; MacDill AFB, Florida; and Los Angeles MEPS, California. After leaving active duty in 1998, he joined the 290th Joint Communications Support Squadron, Florida Air National Guard at MacDill AFB. Following three years in the Florida ANG, he was activated for two years in September 2001 after the events of 9/11. In 2005, he joined the 482nd Communications Squadron at Homestead Air Reserve Base, Florida. He later completed three years of active-duty tours with U.S. Special Operations Command and U.S. Central Command as a Senior Master Sergeant. During his reserve time and before his active-duty tours, he worked as an IT Manager for HQ, U.S. Central Command with Lockheed Martin IT Services and as an IT Training Manager for HQ, U.S. Special Operations Command with Eagan McAllister, Inc., training active-duty special forces communications personnel worldwide as a civilian contractor. Dr. Rod was selected as Chief Master Sergeant (E-9) by the Air Force Reserve Command (AFRC) at Robins AFB, Georgia, and relocated to Keesler AFB, Mississippi, as the Technical Training Reserve Liaison for AFRC, where he managed 5,000 technical school students at 22 different nationwide locations before his retirement.

Honored with an Honorary Doctorate in Ministry for his civic engagement, Dr. Rod is the CEO of Chief Empowerment Network, LLC. This "boutique style" book editing/publishing company aids aspiring authors in writing and marketing their first books and transitioning to paid speaking engagements. Dr. Rod has significantly impacted various sectors through his opioid misuse prevention, emotional wellness programs, and mindset seminars, positively affecting over 40,000 individuals. He had a 2-year contract working closely with the Department of Justice, including the DEA and the FBI. His written works include "**CLIMB**: Face Your Past, Own Your Future," "**CLIMB II**: Time to SOAR," and his recent publication, "**ELEVATE**: Success Workbook for College Students," published in March 2025, which is now being used in colleges across the nation. These works provide deep insights into emotional health and personal development. Dr. Rod's commitment extends beyond his professional achievements, as demonstrated by his selfless service with Omega Psi Phi Fraternity, Inc., his service on various boards, and his advocacy for veterans through his weekly podcast.

Valerie Cunningham

An embodiment of transformation and empowerment, Valerie Cunningham has spent three decades nurturing her spiritual development and mastering the art of mindset coaching. A life-long entrepreneur, graduate of the Master Pastor Internship Program, and the visionary behind Women Living W.E.L.L., Valerie's initiatives focus on empowering individuals to excel in mind, body, and spirit. Her seminars and books, including the "CLIMB" series and the upcoming "Act Like a Queen and Wear Your Crown," inspire profound personal growth and emotional wellness. As a certified Mental Health First Aid instructor, Valerie's reach extends into community mental health advocacy, helping to foster resilience and self-love among diverse populations.

Together on Stage

Dr. Rod and Valerie frequently share the stage, speaking on mental/emotional wellness, mindset, and self-care at esteemed venues like Johns Hopkins Hospital's Pediatric Brain Symposium, Northwestern Mutual Financial, and Miami-Dade College, among others. Their collaborative efforts have led to the development of impactful seminars that resonate deeply with audiences across educational, corporate, military, and healthcare settings.

As co-authors, their synergy brings a unique depth to their writing, blending practical advice with spiritual wisdom to cater to a wide audience eager to improve their lives holistically.

A Legacy of Impact

Together, they stand as a beacon of hope and a source of inspiration, advocating for better emotional/financial health for individuals and communities. Their work demonstrates that with the right mindset and support, transformation is within everyone's reach.

Invite the Cunninghams

Inviting Dr. Roderick and Valerie Cunningham to speak at your event promises an engaging, enlightening, and transformative experience. Their talks are not just speeches but journeys into the essence of human potential and resilience. Whether addressing veterans or seasoned professionals, their message of hope, strength, and renewal is universal.

Introduction

From Service to Civilian: Your New Mission Begins

Welcome to ELEVATE for VETERANS, a guide specifically designed to support your journey from military service to civilian life. Transitioning from the structured, mission-oriented environment of the military to the often less defined civilian world presents unique challenges and opportunities. This workbook focuses on cultivating a balanced, successful post-military life by building on the strengths, skills, and values you developed during your service.

Why "ELEVATE for VETERANS"?

The essence of "ELEVATE for VETERANS" is to lift you to greater heights by enhancing your understanding and implementing key concepts that contribute to a fulfilling post-military life. Each chapter is dedicated to unraveling complex ideas into actionable insights that you can apply directly to your civilian transition, relationships, and career aspirations.

Just as you were trained for military service, this workbook aims to train you for civilian success, with tactics, strategies, and exercises tailored to your unique experiences as a veteran.

Navigating Through "ELEVATE for VETERANS"

This book is structured into three main sections, each addressing critical areas of development:

Personal Growth and Emotional Resilience: These chapters explore the concept of the "Emotional Bucket" as it applies specifically to veterans, acknowledging the unique emotional loads carried by those who have served. From managing service-related stress to repurposing the warrior mindset for civilian challenges, you'll learn how to navigate life's terrain with the same strength and grace that defined your military service.

Financial Wellness and Relationships: We cover veteran-specific financial challenges and opportunities, from maximizing VA benefits to creating post-service financial security. We provide detailed guidance on navigating the VA compensation process to ensure you receive all the benefits you've earned. Additionally, we address the critical transitions in relationships - from military unit cohesion to civilian networking, and from military family dynamics to post-service family reintegration.

Professional Success and Leadership: The final chapters focus on transferring your military leadership, discipline, and skills to civilian success. Whether through entrepreneurship, corporate careers, or public service, you'll learn how to communicate your value, set meaningful post-service goals, and continue making an impact in new environments.

The Tools You Will Gain

Through "ELEVATE for VETERANS," you will be equipped with tools such as post-military goal-setting techniques, transition resilience exercises, civilian relationship-building strategies, and skill translation frameworks; all designed to empower you to make informed decisions for your post-service life. Each chapter includes practical exercises, reflective questions, and affirmations tailored explicitly to the veteran experience.

For Whom Is This Book?

Whether you are a recently discharged veteran facing the immediate challenges of transition, a veteran who has been in civilian life for years but still feels out of place, or an active-duty service member preparing for future transition, "ELEVATE for VETERANS" offers valuable lessons and insights. It's for those who are ready to apply the same dedication they brought to their military service to building a fulfilling civilian life.

As you turn the pages, I invite you to engage fully with the exercises and reflections provided. "ELEVATE for VETERANS" is not just a book to be read - it's a mission to be accomplished. By the end, you'll not only have gained knowledge and skills but also developed a deeper understanding of yourself and your path to post-military success.

Let's begin this transformative journey together. Elevate your mind, elevate your life, and rise to the heights you've always been capable of reaching.

SECTION I

Personal Growth and Emotional Resilience

Understanding Your Emotional Bucket as a Veteran

Chapter

1

Welcome to the beginning of your journey towards emotional wellness in civilian life. In this chapter, we introduce a crucial concept, the "Emotional Bucket." This metaphorical bucket represents your capacity to hold all the emotional responses you've accumulated through various experiences over your entire life, from early childhood through your military service and into civilian transition. Understanding this concept is the first step toward recognizing how both your pre-military and military experiences influence your present emotional state.

From early childhood, we begin to download life's instructions, forming our initial understanding of the world around us. These instructions are not inborn but are deeply influenced by our environments and the key figures in our lives, parents, teachers, family members, and grandparents. They shape our perceptions, beliefs, and emotional responses. As we grow, this foundational programming can deeply affect how we view ourselves and interact with others, and it follows us into military service.

When we experience significant emotional events such as parental abandonment, loss, trauma, or rejection during childhood and adolescence, these are not just mere memories; they leave a lasting imprint. The pain from these experiences is downloaded into our Emotional Bucket, often rewiring our minds and emotional frameworks in profound ways. For many future service members, these early experiences may have already filled a significant portion of their Emotional Bucket before they ever put on a uniform.

The Pre-Military Foundation

It's important to recognize that many individuals enter military service with Emotional Buckets that are already partially filled from childhood experiences. These early experiences can include:

- Parental abandonment through divorce, separation, incarceration, or absence
- Loss of loved ones, including family members, friends, or pets
- Trauma from physical, sexual, or emotional abuse, or life-threatening situations
- Rejection from romantic heartbreaks, social exclusion, or family disappointments

These pre-existing emotional loads often influence behaviors during military service. Veterans may recognize that some of their military-era coping mechanisms, such as drinking, smoking, isolating, fighting, gambling, excessive sleeping, sugar addiction, or shopping addiction, may have roots in these earlier unresolved emotional experiences.

Military Service Adds New Layers

Military service then adds its own unique emotional experiences to an already established foundation. The structured, high-stress environment of military life creates additional opportunities for significant emotional events:

- Combat and operational stress from deployments and high-intensity operations
- Loss of battle buddies and the unique grief that comes with military losses
- Military-specific rejection through comparison with higher-performing soldiers, unfair punishment, or obvious favoritism from leadership
- Identity challenges as military culture shapes and sometimes conflicts with pre-existing beliefs and values
- Moral injury from witnessing or participating in events that conflict with personal values

The military environment, while providing structure and purpose, can also trigger unresolved emotions from one's childhood. For example, a service member who experienced parental abandonment might be particularly affected by leadership that seems uncaring or absent. Someone who faced rejection in childhood may be more susceptible to feeling overlooked or passed over for promotions or opportunities.

The Combined Impact

These "combined" emotional childhood experiences plus military experiences influence who you are now as you transition to civilian life. It shapes your thoughts, reactions, and behaviors in ways that can be both positive and challenging. For many veterans, the emotional bucket that was already partially filled before service became even heavier during military years, creating a significant load to carry into civilian transition.

The coping mechanisms developed during service, some healthy, some not, were often adaptations to manage this combined emotional weight. Understanding that both your pre-military and military experiences contribute to your current emotional state is crucial for effective civilian transition and emotional wellness.

To manage the contents of your Emotional Bucket effectively, it's important to develop strategies that address both your childhood foundation and your military experiences. This involves revisiting these experiences with new tools, such as veteran-focused therapy, mindfulness, and emotional intelligence education, to understand how pre-military and military experiences interact and influence your civilian transition. By doing so, you can lighten the load in your Emotional Bucket, making room for new positive civilian experiences and responses, ultimately leading to a healthier, more balanced emotional life beyond your military service.

The Concept of the Emotional Bucket

Imagine you have an invisible bucket that you carry around inside you. This bucket has been filling throughout your entire life, from early childhood disappointments and traumas, through your military service experiences, and into your civilian transition. Events that add to this bucket can range from minor childhood disappointments to major military traumas. Without proper management, the bucket can overflow, leading to overwhelming feelings that might manifest as physical symptoms, mental health challenges, or difficulty adjusting to civilian environments.

Throughout your military service, this bucket continued to fill with both the weight of unresolved childhood issues and new military-specific experiences. Some experiences sit at the bottom of your bucket, foundational childhood experiences that are heavy and deeply embedded. Others float near the top, more recent military experiences that are easily accessible. Some were processed and contained, while others may have been hastily stashed away during high-stress operations or deployments when there wasn't time to process them properly.

4 Areas that Cause Emotional Pain or Emotional Decline®

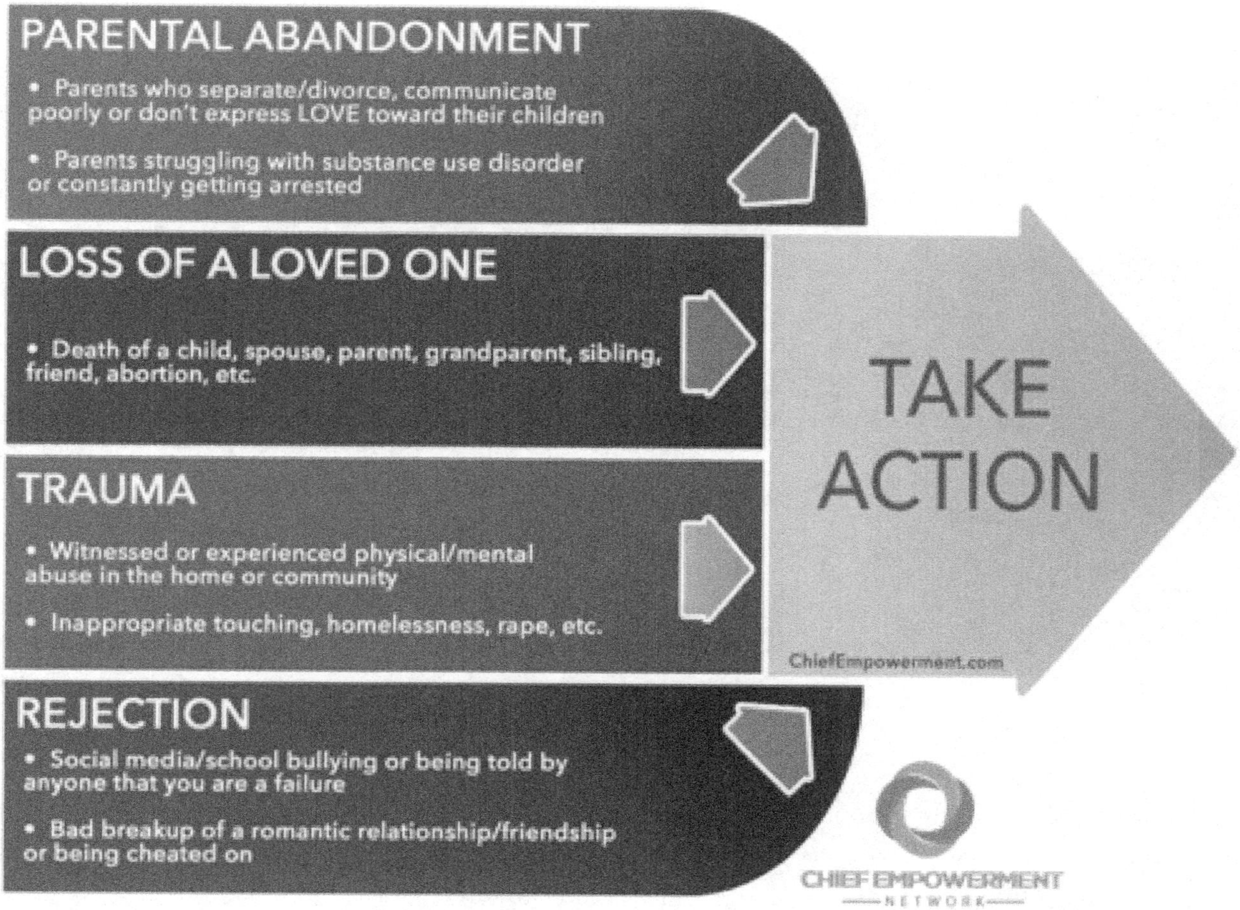

Roderick Cunningham®

WARNING:
Unresolved, any one of these areas CAN lead to a lifetime of high stress, anger, sadness, isolation, anxiety, or depression.

Types of Significant Emotional Events that Fill Your Emotional Bucket

1. **Parental Abandonment**: Whether through divorce, separation, incarceration, or having never been present, the absence of a parent can leave a significant mark. This abandonment can affect trust in relationships and self-esteem, steadily filling your emotional bucket over time.

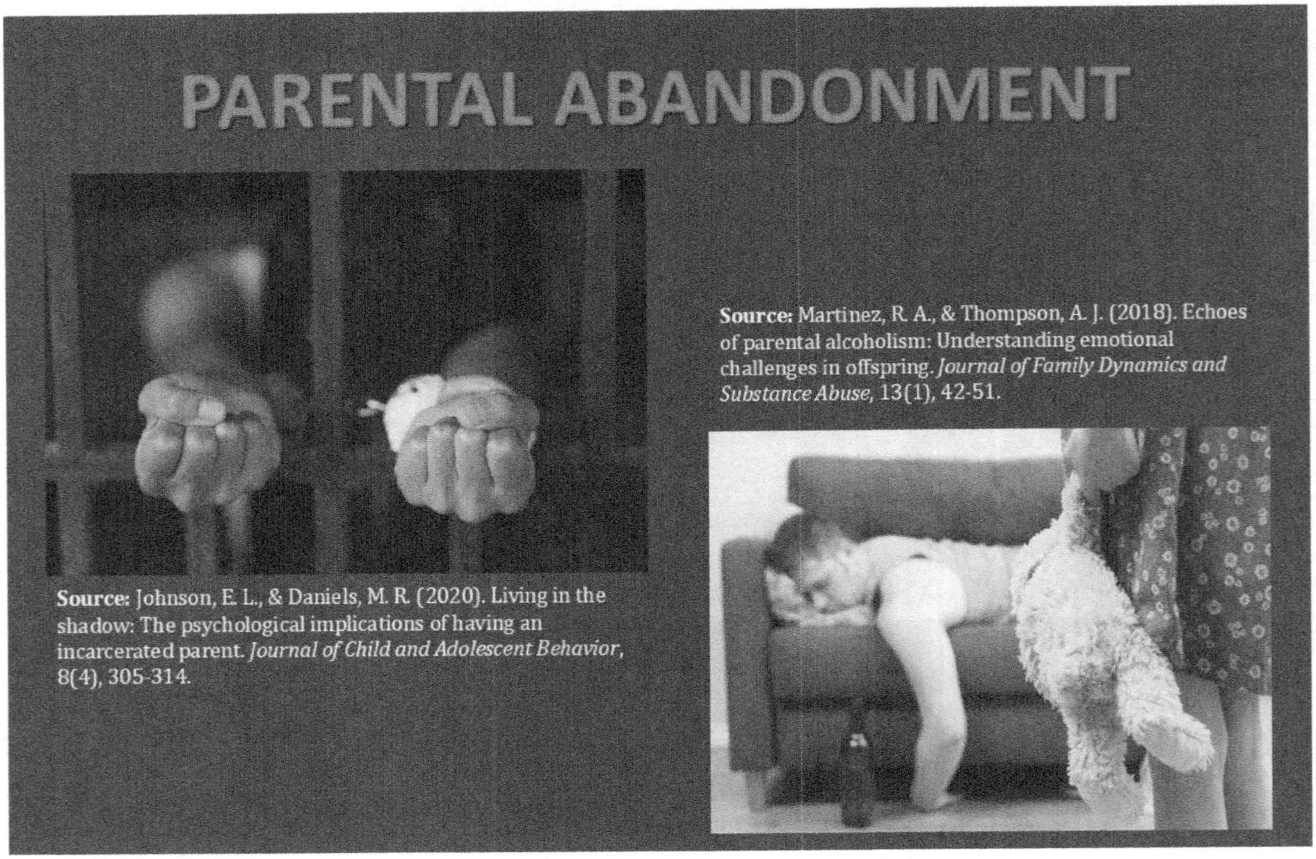

2. **Loss of a Loved One**: Experiencing the death of someone close to you, whether a family member, friend, or pet, contributes to the bucket. Each loss adds a layer of grief that can persist and complicate your emotional health if not addressed. The most significant heartbreaking loss for a child is losing a biological parent. The following is the most significant heartbreaking loss for an adult:

    - Loss of a child
    - Loss of a significant other
    - Loss of a parent
    - Loss of a best friend, mentor, sibling, aunt, or uncle

Source: Harrison, D. L., & Greene, A. L. (2022). Stored Grief: How losses, both human and animal, imprint on our emotional memory. *Journal of Emotion Regulation and Processing*, 7(1), 58-67.

3. **Trauma**: This can include physical, sexual, or emotional abuse, as well as any life-threatening situations you've endured. Trauma can have a profound and lasting impact, continually adding to the bucket each time memories resurface or similar situations occur.

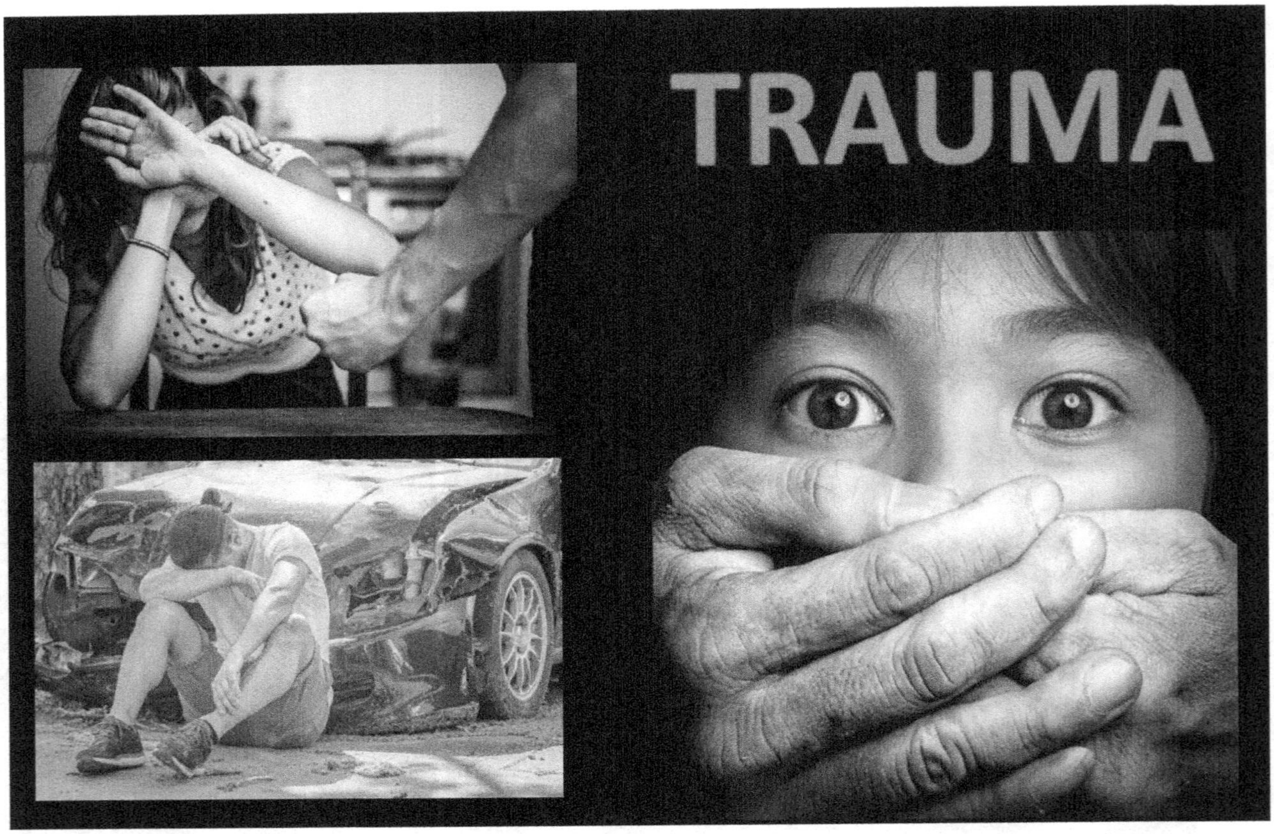

4. **Rejection**: From romantic heartbreaks to social exclusion or even professional setbacks like being fired, rejection strikes at the core of our need for belonging and self-worth, adding to the emotional load we carry.

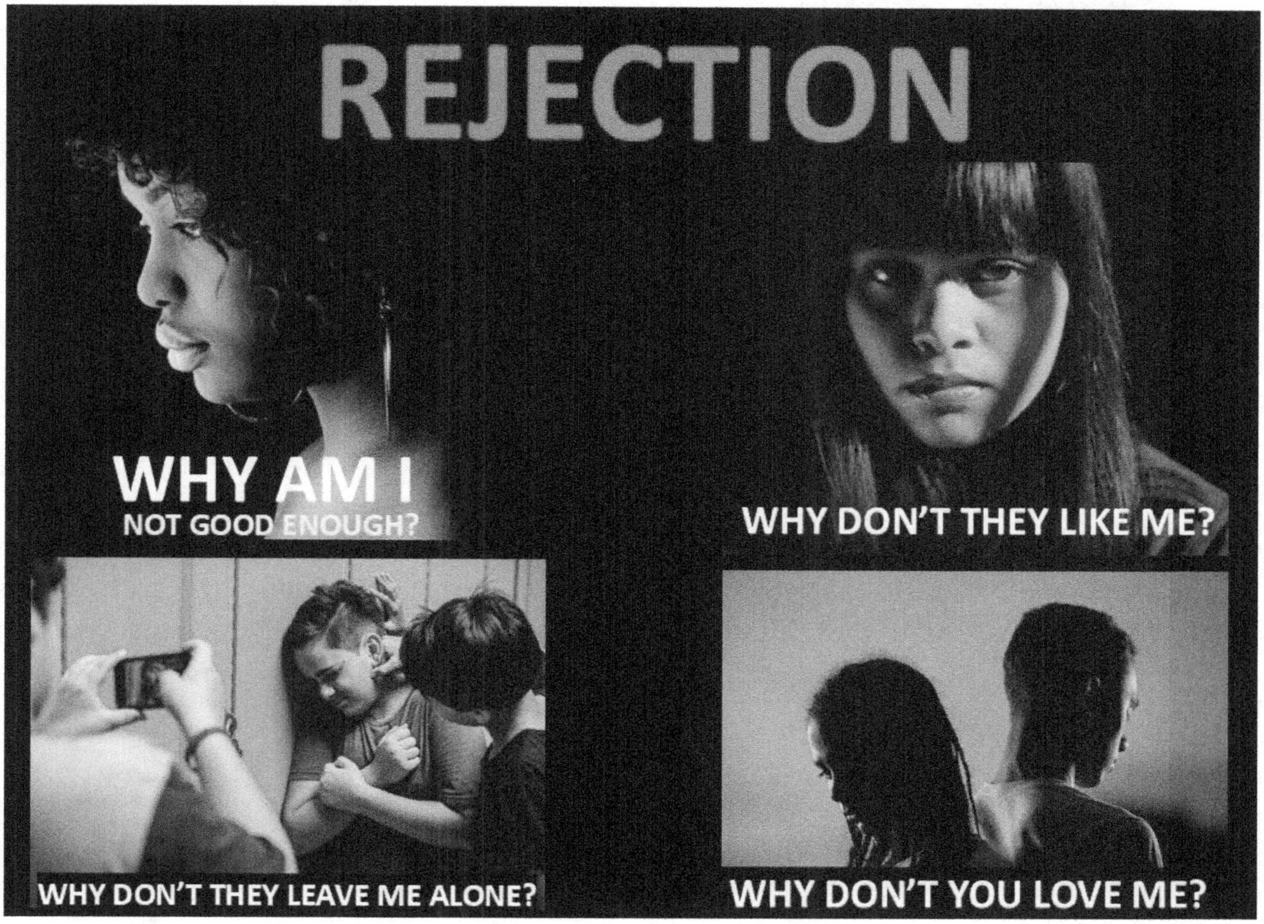

Military-Specific Events:

1. **Combat and Operational Stress**: High-intensity operations, combat exposure, and deployment stress add significant weight to buckets that may already be partially filled from childhood experiences.

2. **Loss of Battle Buddies**: The deep bonds formed in military service make the loss of comrades particularly significant, especially for those who have already experienced early losses.

3. **Military Rejection and Comparison**: Being compared unfavorably to other service members, experiencing obvious favoritism from leadership, or receiving unfair punishment can trigger childhood rejection experiences, making the impact even more severe.

4. **Family Rejection Due to Deployments and Your Mental State**: The operational tempo of multiple deployments can be tough on a romantic relationship and missing milestones with children (e.g., sporting events, graduations, recitals, birthdays, holidays, etc.). When you do return, you may be easily agitated, which may scare family members.

5. **Military Transition Shock**: The contrast between military structure and civilian ambiguity creates its own stress, particularly challenging for those whose childhood lacked stability.

How the Combined Emotional Bucket Affects You

As your Emotional Bucket fills with both childhood and military experiences, it becomes heavier and harder to carry, especially during civilian transition. This weight can manifest as hypervigilance in safe environments (triggered by both childhood trauma and military training), isolation from civilians who "don't understand," anger disproportionate to civilian situations, or emotional numbness. The fuller the bucket, the less room you have to manage new civilian stressors effectively.

Imagine your Emotional Bucket was already partially filled with childhood experiences, early grief, family dysfunction, or childhood trauma. Then military service added combat experiences, military losses, and the stress of high-threat environments. Each of these elements represents significant emotional content that you've been carrying, possibly for many years. Now, when a new challenge, a "civilian brick", such as job search difficulties, relationship misunderstandings, or feeling out of place, is added, it doesn't just add to the bucket; it can cause a critical overflow.

This new civilian challenge can be particularly destabilizing because it doesn't enter an empty bucket; it falls into one that's been filling since childhood and was further loaded during military service. The impact of this new stressor is much more severe, as it doesn't find the space to be processed in the usual ways. This helps explain why seemingly "minor" civilian issues can sometimes trigger disproportionate responses in veterans; the new challenge isn't being processed in isolation, but on top of a lifetime of accumulated emotional experiences.

The coping mechanisms you developed during service, whether healthy or unhealthy, were often attempts to manage this combined emotional load. Recognizing patterns like drinking, smoking, isolating, fighting, gambling, excessive sleeping, or addictive behaviors as responses to emotional overload (rather than character flaws) is crucial for effective civilian transition.

It's essential to develop effective strategies for managing the contents of your Emotional Bucket during transition and beyond, addressing both childhood foundations and military experiences. Proactive emotional management involves regular self-care practices like veteran-focused therapy, mindfulness, physical exercise, and healthy social interactions, which can help drain the accumulated emotions from both life phases. Additionally, learning and practicing resilience-building techniques such as cognitive restructuring and stress reduction can prepare you to handle new civilian challenges more effectively.

We must prioritize emptying our Emotional Bucket not only to cope with the present but also to prepare for civilian life. By keeping the bucket from becoming too full, we ensure there is always enough capacity to manage new challenges without being overwhelmed. This proactive approach is essential in maintaining emotional stability and safeguarding mental health, especially when navigating through the inevitable storms of civilian transition while carrying the combined weight of a lifetime of experiences.

Exercise

UNDERSTANDING YOUR EMOTIONAL BUCKET AS A VETERAN

Objective: This exercise aims to help you visualize and manage the contents of your Emotional Bucket, encouraging you to identify significant emotional events from both your pre-military life and military service, and understand their combined impact on your emotional wellness in civilian life.

Materials Needed:

- Pen and paper or a digital journal
- A quiet, comfortable space for reflection

Exercise Description:

1. **Mapping Your Complete Emotional Bucket:**
    - Draw a large bucket on your paper or digital canvas. This represents your Emotional Bucket.
    - Divide your reflection into two phases: Pre-Military Life (childhood through enlistment) and Military Service (enlistment through discharge).
    - For Pre-Military: Identify events that had a significant emotional impact, such as parental abandonment, loss, trauma, or rejection.
    - For Military Service: Identify service-related events including combat experiences, loss of comrades, rape/sexual harassment, military rejection/ comparison, non-judicial punishment, and other challenging experiences like an unfaithful spouse or divorce.
    - For each event, draw a 'brick' inside your bucket. Use different colors or patterns to distinguish pre-military (foundation bricks) from military (upper-level bricks). Label each brick with the event and associated emotion.
2. **Analyzing the Combined Load:**
    - Look at the bricks you've placed in your bucket. Which pre-military experiences seem to have influenced your military service experience?
    - Identify any patterns where military experiences triggered or amplified childhood emotions.

- Note which experiences feel heaviest and why they might carry more emotional weight.

3. **Behavioral Pattern Recognition:**
   - Reflect on any unhealthy coping mechanisms you developed during service (drinking, smoking, isolating, fighting, gambling, excessive sleeping, shopping, partying, cheating, sexual addiction, food or sugar addiction, etc.).
   - Consider how these behaviors might have been attempts to manage the combined weight of childhood and military experiences.
   - Write down your observations about these patterns without judgment.

4. **Emotional Impact Reflection:**
   - Reflect on how the combination of childhood and military experiences has shaped your personality, relationships, and life choices as you transition to civilian life.
   - Consider how childhood experiences may have made certain military experiences more difficult.
   - Note any patterns or triggers that emerge from this combined load and how they affect your current civilian transition.

5. **Developing Comprehensive Coping Strategies:**
   - For each major brick (both childhood and military), brainstorm coping strategies that could lighten its emotional load. These might include veteran-specific therapy, trauma-informed counseling, talking to fellow veterans, engaging in physical activity (self-care), or practicing mindfulness.
   - Prioritize strategies that address both childhood foundation and military experiences.
   - Decide on one strategy to implement immediately that addresses your most significant combined emotional burden.

6. **Preventative Measures for Civilian Life:**
   - Reflect on ways to prevent new civilian experiences from adding excessive weight to your already-loaded bucket.

- Consider how understanding your pre-military and military emotional patterns can help you navigate civilian challenges more effectively.
- Write down at least two preventative measures specific to your transition needs.

7. **Sharing Insights:**
    - Optionally, share your complete Emotional Bucket map with a trusted fellow veteran, a counselor familiar with military culture, or in a veteran support group.
    - Discussing your bucket's complete history can provide insights and support from others who understand both the challenges of difficult childhoods and military service.

8. **Continuous Evaluation:**
    - Revisit your Emotional Bucket map every few months during your civilian transition. Update it as you process old emotions from both life phases and encounter new civilian challenges.
    - Track how addressing childhood foundations impacts your ability to process military experiences and vice versa.

Reflect on this exercise's impact on your understanding of yourself and your emotional health as a veteran in transition. Write a brief reflection on how visualizing your complete Emotional Bucket, including both pre-military and military experiences, has changed your perception of your transition challenges and what steps you feel are most crucial for your wellness journey going forward.

By engaging in this exercise, you take an important step towards not just understanding but actively managing your emotional health as a whole person, not just as a veteran. This comprehensive approach is essential in maintaining a balanced emotional state and ensuring that you are equipped to handle whatever new challenges civilian life may bring while honoring the full scope of your life experiences.

Self-Reflective Questions

1. Reflect on the major events in both your childhood and military service that have contributed to your Emotional Bucket. How do you think your pre-military experiences influenced your military service?

2. Consider the emotions associated with both childhood and military events. Are there patterns where similar types of experiences (like rejection or loss) occurred in both phases of your life?

3. How do the combined childhood and military experiences affect your current civilian relationships and decision-making?

4. Can you identify any unhealthy coping mechanisms you developed during service that might have roots in earlier childhood experiences?

5. What steps have you taken to address both your childhood foundation and military experiences, and how might a comprehensive approach be more effective for your civilian transition?

Affirmations

1. I acknowledge both my childhood and military experiences and embrace the journey towards complete healing in civilian life.
2. I am more than the sum of my difficult experiences; I am a resilient individual capable of growth and change beyond both my childhood and military past.
3. I choose to release the emotions from my entire life that no longer serve my well-being in civilian transition.

Actionable Steps

1. Begin by journaling about both childhood and military events you identified as filling your Emotional Bucket, focusing on how they might be connected and how they affect your civilian life.
2. Seek therapy or counseling that addresses both childhood trauma and military transition, preferably from someone trained in both areas.
3. Engage in mindfulness exercises that help you observe and process emotions from your complete life experience without judgment.
4. Create a comprehensive emotional wellness plan that addresses both your childhood foundation and military experiences through daily or weekly activities.
5. Connect with veteran support groups or programs that understand the complexity of carrying both childhood and military emotional experiences into civilian life.

Understanding your complete Emotional Bucket, including both childhood and military contributions, is the first step towards achieving emotional wellness in civilian life. By recognizing how your entire life experience has shaped your emotional landscape, you can begin to take proactive steps to manage and reduce this burden comprehensively. Remember, the goal is to prevent your bucket from overflowing by learning from all your experiences and discovering who you can become in civilian life with a fuller understanding of your complete emotional history.

Evidence-Based Research: Research suggests that recognizing and naming emotions connected to specific life experiences, both childhood and military, can significantly reduce the intensity of these emotions, leading to better emotional management during civilian transition (Gross, 1998). Studies with combat veterans have shown that those who address both their pre-military and military emotional experiences show greater resilience and more successful adaptation to civilian life, particularly when treatment approaches acknowledge the interplay between childhood and military trauma.

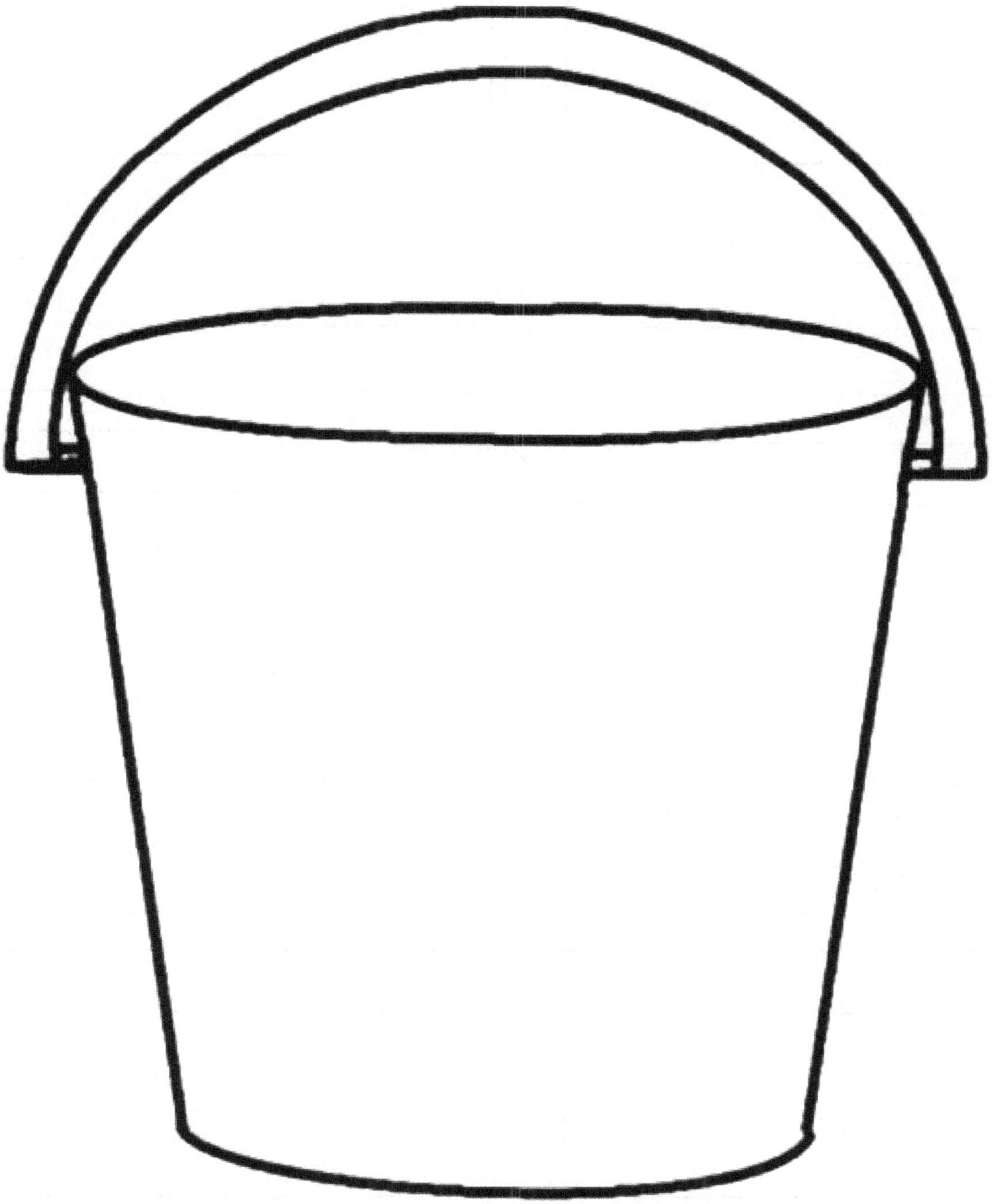

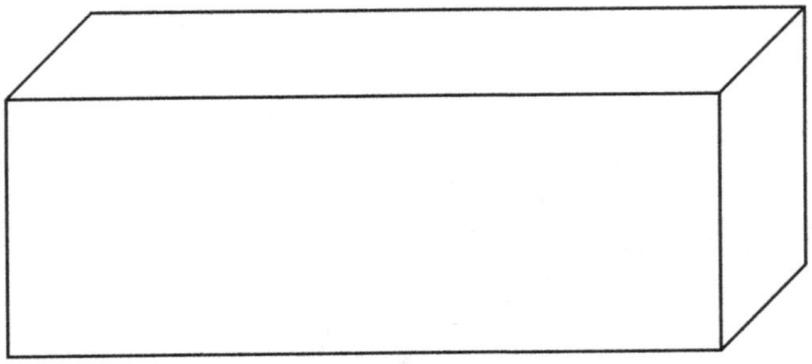
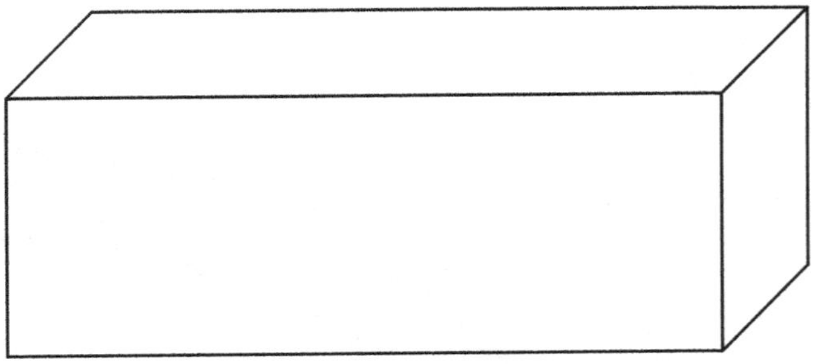
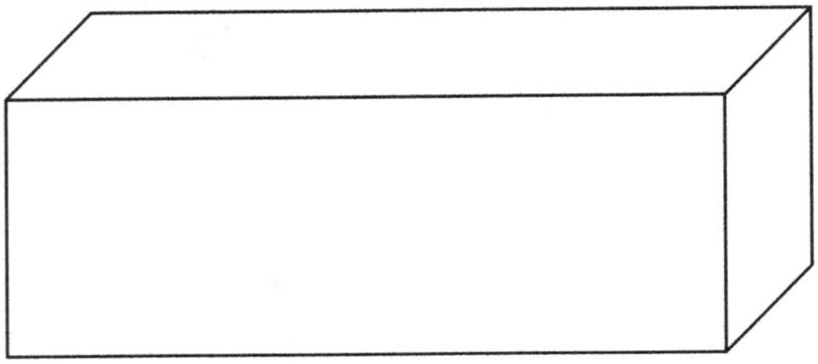
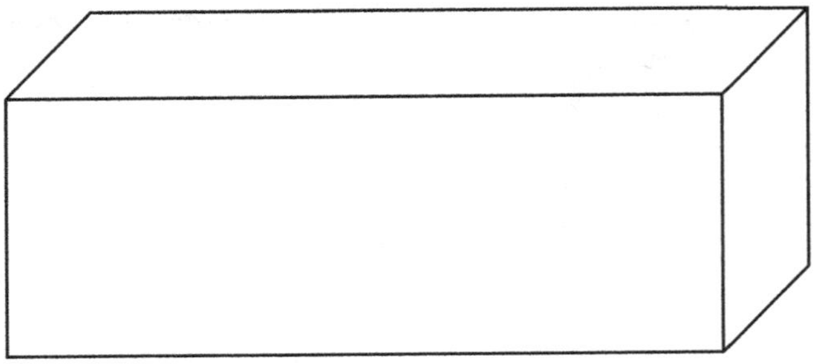
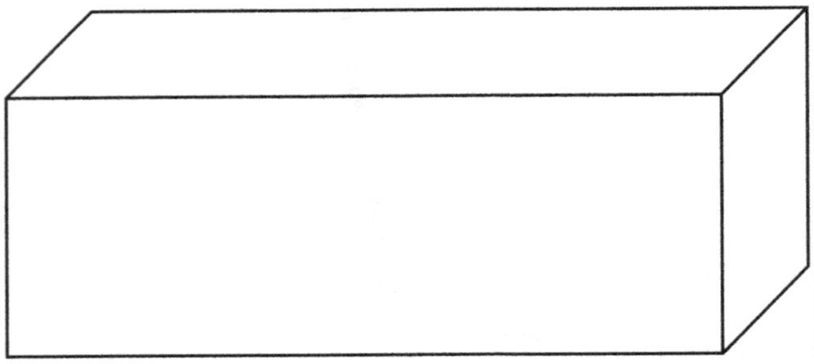
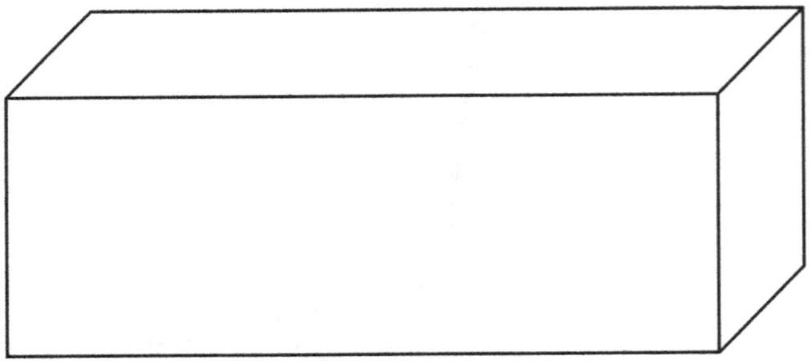

NOTES

The Five Spigots: Draining Your Emotional Bucket

Chapter

2

Now that you understand the concept of the Emotional Bucket and recognize the types of experiences that contribute to its fill, this chapter introduces you to the five key methods, or "spigots," that can help you drain this bucket effectively during and after your military-to-civilian transition. These spigots are not quick fixes but are practices that, when incorporated into your post-military life, can help manage and mitigate the emotional load you carry from your service.

5 Steps to Emotional Wellness

The Five Spigots Explained

1. **Spirituality**: This spigot taps into the intrinsic need for purpose and connection that many veterans feel acutely after leaving the structured mission-oriented environment of military service. Engaging in spiritual practices - whether through traditional religious avenues, mindfulness meditation, or connecting with nature - offers profound solace and stability amidst the emotional turmoil of transition. Many veterans find that spirituality helps them make meaning of their service experiences and provides a framework for understanding their new civilian mission. Connecting with a higher power or purpose often elevates a veteran's state of being, manifesting as increased peace and spiritual grounding - qualities that enhance overall well-being during transition challenges.

    - **Actionable Steps**:
        - Identify spiritual practices that resonated with you during service (whether formal religious observances or informal moments of connection) and adapt them to civilian life.
        - Connect with veteran-friendly spiritual communities that understand military experiences.
        - Develop a regular spiritual routine that provides structure similar to what you valued in military life.
    - **Affirmations:**
        - I am connected to a purpose greater than my military service.
        - My spiritual practices bring me peace and clarity in civilian life.
        - I am guided through my transition challenges by something greater than myself.

2. **Forgiveness**: Military service often creates complex situations requiring difficult decisions under extreme pressure. Many veterans carry guilt, regret, or resentment - whether toward themselves, leadership, the enemy, or the civilian world that seems to have moved on without them. This emotional weight occupies a substantial amount of space in your Emotional Bucket. Forgiveness in this context doesn't mean excusing or forgetting but instead releasing the emotional hold these experiences maintain over you, allowing you to operate more effectively in civilian life.

- **Actionable Steps**:
  - Identify specific service-related experiences where forgiveness (of self or others) could drain your emotional bucket.
  - Practice empathy by understanding the context and limitations under which decisions were made during high-stress military situations.
  - Engage in forgiveness rituals adapted to military experiences, such as writing letters to fallen comrades or to your younger military self.
- **Affirmations:**
  - I choose to forgive to free myself from the burden of service-related resentments.
  - Letting go of past military burdens allows me to embrace my civilian potential.
  - Forgiveness brings me closer to true peace of mind beyond my service identity.

3. **Counseling**: Professional help from counselors or therapists who understand military culture and transition challenges can be crucial in managing your Emotional Bucket. VA-affiliated or military-informed mental health professionals can provide strategies and tools to process service-related experiences in ways that enhance rather than hinder your civilian adaptation.

- **Actionable Steps**:
    - Research therapists who specialize in veteran transitions, military trauma, or military cultural competence.
    - Schedule an initial consultation with a VA mental health provider or a community provider who works with veterans.
    - Commit to regular sessions as part of your civilian transition strategy, viewing it as an operational necessity rather than a weakness.
- **Affirmations:**
    - I am tactically smart to seek support from those who understand military experiences.
    - Therapy is a mission-critical space for me to process my service experiences.
    - Each counseling session improves my operational effectiveness in civilian life.

4. **Purpose**: Finding a meaningful civilian purpose involves understanding what truly matters to you beyond your military identity. This pursuit can redirect your energy from past service-related pains to positive civilian action, providing a sense of accomplishment and satisfaction that gradually displaces the emotional residue in your bucket. Many veterans find purpose in continued service through different means - community involvement, veteran advocacy, or careers that utilize their military-honed skills for different missions.

   - **Actionable Steps**:
     - Identify aspects of your military service that gave you the greatest sense of meaning and consider civilian contexts where similar values apply.
     - Engage in structured purpose-finding activities designed for veterans, such as community service or veteran mentorship programs.
     - Volunteer for causes that connect to values you honored during service, strengthening the bridge between your military past and civilian future.
   - **Affirmations:**
     - My sense of purpose continues beyond my military service.
     - Every civilian action aligned with my core values honors my service in a new way.
     - My mission continues in new forms that serve both myself and others.

5. **Journaling and Creative Expression**: Military culture often emphasizes the containment of emotions for operational effectiveness. While functional during service, this pattern can become problematic in civilian life. Expressive practices like journaling, creating art, or participating in veteran writing programs allow you to drain emotions from your bucket that may have been necessarily suppressed during service. These activities provide a safe outlet for expressing service-related feelings, leading to significant reductions in emotional stress and improved civilian adaptation.

   - **Actionable Steps**:
     - Keep a transition journal documenting your journey from military to civilian identity.
     - Explore veteran-focused creative arts programs that provide structured ways to express service experiences.
     - Consider storytelling, whether written or verbal, as a way to integrate your military experiences into your complete life narrative.
   - **Affirmations:**
     - My military experiences become wisdom when I express and process them.
     - Through creative expression, I transform challenging service memories into strength.
     - Sharing my military story helps me and potentially serves others who walk this path.

Implementing these spigots requires commitment and patience, but they are powerful tools for draining your Emotional Bucket. As you integrate these practices into your civilian life, you'll find that they not only help empty your service-related emotional load but also enrich your overall transition experience, leading to greater emotional health and fulfillment beyond your military service.

Evidence-Based Research: Studies specifically focusing on veteran populations have shown that spirituality, structured processing of service experiences, and finding new forms of purpose significantly lower transition stress and enhance overall emotional well-being after military service (Kopacz & Connery, 2015).

Exercise

IMPLEMENTING THE FIVE SPIGOTS TO DRAIN YOUR VETERAN EMOTIONAL BUCKET

Objective: This exercise is designed to help veterans understand and apply the Five Spigots method to drain their Emotional Bucket. Through guided activities, reflections, and affirmations, veterans will learn practical ways to reduce service-related emotional overload and enhance their emotional well-being during civilian transition.

Instructions: Each veteran will undertake activities related to each of the five spigots over a period of one week. You are encouraged to keep a detailed journal of your experiences, thoughts, and feelings throughout the exercise.

Activities:

1. Spirituality

- **Daily Activity:** Dedicate 15 minutes each morning to engage in spiritual practices of your choice (meditation, prayer, reading spiritual texts, Bible passage reading, or simply being in nature).
- **Weekly Task:** Attend at least one service or meeting that aligns with your spiritual values, preferably one that includes other veterans or addresses veteran experiences.
- **Reflection:** At the end of the week, write a reflection on how these practices affected your emotional state and sense of connection to something beyond your military identity.

2. Forgiveness

- **Daily Activity:** Identify one aspect of your military experience that you hold resentment about (toward yourself, others, or the system) and write down how this has affected your civilian transition.
- **Weekly Task:** Write a forgiveness letter addressing this aspect (no need to send it) to release some of its emotional weight from your bucket symbolically. Be sure to burn the letter in the kitchen sink.
- **Reflection:** Reflect on the process of forgiveness and any changes in your emotional state post-activity, particularly noting if civilian interactions feel different afterward.

3. Veteran-Specific Counseling

- **Research Task:** Spend some time researching therapists or counselors who specialize in working with veterans in your area, including VA options and community providers.
- **Initial Step:** Make an appointment for a consultation with a counselor who has military cultural competence.
- **Weekly Reflection:** After the consultation, journal about your experience and feelings towards continuing therapy as part of your transition strategy.

4. Purpose

- **Daily Activity:** Write down three potential civilian purposes that might be meaningful to you, connecting them to values you honored during service.
- **Weekly Task:** Engage in at least one activity that aligns with these potential purposes, whether volunteer work, informational interviews, or community engagement.
- **Reflection:** At the end of the week, reflect on how pursuing these purpose-aligned activities impacted your sense of civilian identity and emotional wellbeing.

5. Journaling, Art, and Expressive Processing

- **Daily Activity:** Spend at least 10 minutes each day journaling, creating art, or recording spoken reflections about your transition experience.
- **Weekly Task:** Consider sharing one piece of your expressive work with a trusted person or veteran peer group, or visit a veteran arts program in your community.
- **Reflection:** Reflect on how these expressive activities have helped drain service-related emotions from your bucket over the week.

Submission Requirements: At the end of the week, create a comprehensive report that includes:

- **Journal Entries:** Daily entries that document your activities, thoughts, and emotional changes.
- **Reflections:** Detailed reflections on each spigot and how the activities influenced your emotional health and transition experience.
- **Affirmations:** Write a personal affirmation that you found most impactful during the week.

Assessment: Your submission will be assessed based on:

- **Completeness:** All tasks and reflections are thoroughly completed.
- **Insightfulness:** Depth of insights gained and ability to connect activities with emotional changes.
- **Personal Growth:** Evidence of personal growth or changes in emotional management strategies specific to your transition challenges.

This exercise aims to provide veterans with practical tools and insights into managing their emotional health during transition, fostering resilience, and enhancing overall well-being. By actively participating in these activities, veterans can begin to drain their Emotional Buckets and cultivate a healthier emotional landscape in civilian life.

Journaling Space for Weekly Reflection:

Self-Reflective Questions

1. **Self-Awareness Evaluation**: Which of the five spigots (Spirituality for Veterans, Service-Informed Forgiveness, Veteran-Specific Counseling, Post-Service Purpose, Expressive Processing) do you currently use the least, and why might you have avoided it during your transition?

2. **Challenge of Service-Related Forgiveness**: Reflect on a past military situation where forgiveness (of yourself or others) could have served as a spigot to drain your Emotional Bucket. What has held you back from this form of forgiveness, and how might you approach it differently now?

3. **Spirituality and Military-Civilian Connection**: How does your spiritual practice (or lack thereof) impact your emotional well-being during transition? Describe any specific moments where spirituality significantly uplifted you during a civilian challenge or, conversely, a time when you felt a void due to its absence.

4. **Role of Veteran-Informed Counseling**: Have you ever considered professional counseling to drain service-related emotions from your bucket? What perceptions or experiences have shaped your view on seeking help from a mental health professional who understands military culture?

5. **Purpose Beyond the Uniform**: How closely do you believe your daily civilian actions align with your sense of purpose? Can you identify any recent activities that either significantly contributed to or detracted from your sense of continued mission after service? How did these experiences affect your emotional state?

Notes

Journaling Space for Weekly Reflection:

Journaling Space for Weekly Reflection:

Journaling Space for Weekly Reflection:

Journaling Space for Weekly Reflection:

Journaling Space for Weekly Reflection:

Combat Stress to Civilian Stress - Building Your Baseline

Chapter

3

Military service exposes you to unique stressors and demands a specific set of responses. From the controlled stress of basic training to the high-stakes pressure of deployment environments, your military experience has trained you to respond to stress in ways that were essential for mission success and survival. However, as you transition to civilian life, these same stress responses may no longer be appropriate or helpful. This chapter explores how to recognize your military-developed stress baseline and adapt it for civilian success.

Understanding Military vs. Civilian Stress

Military stress is often characterized by:

- **Clear threat identification** - You knew who the enemy was and what to watch for
- **Defined protocols** - There were established procedures for handling most situations
- **Unit support** - Your battle buddies had your back in high-stress situations
- **Mission-focused responses** - Stress responses were directed toward accomplishing the mission
- **Controlled environments** - Even in combat, there were rules of engagement and command structures
- **Rank recognition** - Your authority and experience were immediately visible and respected

Civilian stress, by contrast, is often:

- **Ambiguous** - Threats are less clear and more subjective
- **Unstructured** - There may be no standard operating procedures for handling situations
- **Individually managed** - You're expected to handle stress largely on your own
- **Relationship-focused** - Stress responses must consider multiple stakeholders and social dynamics
- **Open-ended** - Problems may not have clear solutions or timelines
- **Rank-blind** - Your military achievements and rank mean nothing to civilian colleagues

The Reality of Civilian Powerlessness

One of the most stressful aspects of civilian transition is the sudden loss of military authority. In the military, your rank immediately communicated your level of responsibility, experience, and authority. Whether you were a Colonel commanding 20,000 troops or a Sergeant with significant responsibility for your unit, civilians don't care about your military rank or the scope of your previous responsibilities.

This powerlessness can be particularly jarring when you're accustomed to having your authority respected and your orders followed. In civilian workplaces, you must earn respect and influence through different means. Your military experience may be valued, but your former rank carries no weight. This reality can create stress as you navigate:

- **Starting at the bottom** - Despite years of military leadership, you may enter civilian roles at entry or mid-level positions
- **Proving yourself again** - Your track record doesn't automatically transfer; you must demonstrate your value in civilian terms
- **Different leadership styles** - Civilian subordinates cannot be commanded; they must be led with compassion and empathy
- **Collaborative decision-making** - Decisions often require consensus-building rather than command authority

Recognizing Your Military Stress Baseline

Your "stress baseline" refers to the default behaviors you typically revert to under significant stress. During military service, you may have developed both healthy and unhealthy stress responses. Understanding these patterns is crucial for civilian adaptation.

Common Military-Developed Stress Responses:

Healthy Military Responses that Transfer Well:

- **Mission-first mentality** - Focusing on objectives and problem-solving
- **Rapid decision-making** - Making quick decisions with limited information
- **Physical fitness** - Using exercise to manage stress and maintain readiness
- **Team coordination** - Working effectively with others under pressure
- **Preparedness** - Planning for contingencies and maintaining readiness

Military Responses that May Need Adaptation:

- **Hypervigilance** - Constant alertness that may be exhausting in civilian environments
- **Emotional suppression** - Compartmentalizing emotions that may need processing in civilian life
- **Authoritarian communication** - Direct, command-style communication that may not work in civilian relationships
- **High-stress tolerance** - Accepting stress levels that might be unhealthy in civilian contexts
- **All-or-nothing thinking** - Binary thinking that may not suit complex civilian situations
- **Rank-based expectations** - Expecting automatic respect and authority based on military experience

Potentially Problematic Stress Responses:

Veterans may have developed less healthy stress baselines that served as coping mechanisms during service but become problematic in civilian life:

- **Substance use** - Alcohol, tobacco, marijuana, or other substances to manage stress
- **Social isolation** - Withdrawing from civilian relationships and support systems
- **Aggressive responses** - Using intimidation or confrontation to manage civilian conflicts
- **Escape behaviors** - Excessive gaming, shopping, gambling, or other avoidance activities
- **Physical symptoms** - Stress manifesting as headaches, sleep issues, or other health problems
- **Authority frustration** - Becoming angry or resentful when civilian authority isn't automatically granted

Building a Civilian-Appropriate Stress Baseline

The goal is not to eliminate your military-developed strengths but to adapt them for civilian success while developing new, healthier responses to civilian stressors.

1. Assess Your Current Responses:

- **Identify triggers** - What civilian situations cause you the most stress?
- **Recognize patterns** - How do you typically respond when stressed in civilian contexts?
- **Evaluate effectiveness** - Are your current responses helping or hindering your civilian goals?
- **Acknowledge powerlessness** - How do you react when your military experience isn't automatically valued?

2. Adapt Military Strengths:

- **Mission-first to goal-oriented** - Apply your mission focus to civilian career and personal goals
- **Unit cohesion to relationship building** - Use your teamwork skills to build civilian support networks
- **Tactical thinking to strategic planning** - Apply your planning skills to civilian challenges
- **Physical readiness to wellness** - Maintain fitness for health rather than combat readiness
- **Command presence to influential leadership** - Transform authoritative leadership into inspirational leadership

3. Develop New Civilian Skills:

- **Emotional intelligence** - Learning to recognize and express emotions appropriately in civilian contexts
- **Collaborative communication** - Adapting from command-style to collaborative communication
- **Ambiguity tolerance** - Learning to function effectively in unclear or changing situations
- **Stress processing** - Developing healthy ways to process stress rather than just enduring it
- **Influence without authority** - Learning to lead through persuasion, empathy, and relationship-building

Leading Civilians with Compassion and Empathy

One of the most critical adaptations you'll need to make is in your leadership style. Military leadership often relies on rank, discipline, and clear chains of command. Civilian leadership, especially when managing civilian subordinates, requires a fundamentally different approach:

Military Leadership Model:

- Authority derived from rank
- Orders expected to be followed
- Discipline through consequences
- Clear hierarchy and structure
- Mission-first, people-second approach

Civilian Leadership Model:

- Authority earned through competence and relationships
- Collaboration and buy-in required
- Motivation through inspiration and empowerment
- Flatter organizational structures
- People-first approach that considers individual needs and circumstances

Strategies for Leading Civilians:

- **Listen first** - Understand their perspectives and concerns before providing direction
- **Explain the "why"** - Civilians need to understand the reasoning behind decisions
- **Show empathy** - Acknowledge their challenges and demonstrate genuine care
- **Be patient** - Civilians may need more time to understand and buy into new approaches
- **Provide support** - Offer resources and assistance rather than just expectations
- **Recognize contributions** - Acknowledge good work and celebrate successes

Healthy Civilian Stress Responses for Veterans

Self-Care Activities: Engaging in physical activities such as weightlifting, aerobics, running, yoga, biking, walking, dancing, or swimming can significantly reduce stress levels. Other forms of self-care include gardening, listening to music, walking on the beach or in nature, caring for pets or horses, burning scented candles, reading, praying, meditating, visualizing your future self, listening to a spiritual message or motivational speech or podcast on YouTube, and other hobbies that bring joy and relaxation.

Self-Care Tempo: Self-care is something that most people only do twice a week. However, practicing self-care every morning is essential to start your day, as it helps bring out your joy and improves your mood and energy levels.

Routine Development: Establishing a daily routine that includes time for work and relaxation can help prevent stress from becoming overwhelming.

Emotional Processing:

- **Veteran support groups** - Connecting with others who understand military experience
- **Professional counseling** - Working with therapists who understand military culture
- **Journaling** - Processing experiences and emotions through writing
- **Art and Creative Expression** - Art, music, or other creative outlets for emotional expression

Social Connection:

- **Veteran organizations** - Maintaining a connection with the military community while building civilian relationships
- **Mentoring** - Sharing your experience with younger veterans or civilians
- **Community involvement** - Finding civilian communities that align with your values
- **Family time** - Investing in relationships with family members who support your transition

Professional Development:

- **Continuous learning** - Taking courses or developing skills relevant to civilian careers
- **Networking** - Building professional relationships in your chosen civilian field
- **Goal setting** - Applying military planning skills to civilian career advancement
- **Skill translation** - Learning to communicate your military skills in civilian terms
- **Influence development** - Learning to lead and influence without relying on rank or authority

Incorporating the Emotional Bucket Concept

Understanding how your stress baseline relates to your Emotional Bucket is crucial. When your bucket is full, you're more likely to revert to old stress patterns, even unhealthy ones, because they feel familiar and provided some relief in the past. When your bucket is better managed, you have more capacity to practice new, healthier civilian stress responses.

Impact of a Full Emotional Bucket on Stress Response:

- **Reduced flexibility** - Less ability to adapt military responses to civilian contexts
- **Increased reactivity** - More likely to respond with military intensity to civilian stressors
- **Poor decision-making** - Stress responses driven by emotion rather than rational assessment
- **Authority frustration** - More likely to become angry when civilian respect isn't automatically given

Benefits of Managing Your Emotional Bucket:

- **Enhanced adaptability** - Better ability to modify responses based on civilian context
- **Improved relationships** - More appropriate responses that strengthen rather than strain civilian relationships
- **Better health outcomes** - Reduced chronic stress and its physical impacts
- **Increased influence** - Better able to earn respect and authority through civilian-appropriate methods

Exercise

STRESS BASELINE ASSESSMENT AND ADAPTATION

Objective: This exercise helps veterans identify their current stress responses and develop personalized strategies for adapting military stress management to civilian contexts.

Instructions:

1. Military Stress Response Inventory:

- Spend one week tracking every instance where you feel significant stress in civilian situations. Note what triggered the stress, how you reacted, and what the outcome was.
- Categories to track: Work stress, relationship stress, financial stress, social stress, authority/powerlessness stress, and transition-related stress.

2. Pattern Analysis:

- Review your inventory to identify patterns. Ask yourself:
  - Which responses served you well in the military but may not work in civilian life?
  - Which civilian situations trigger military-style responses?
  - When do you feel most frustrated by lack of automatic authority or respect?
  - What healthy military responses could be adapted for civilian use?
  - What new responses might you need to develop?

3. Response Classification:

- Classify your stress responses into:
  - **Keep and adapt** - Military responses that work well but may need modification
  - **Replace** - Military responses that don't serve you in civilian contexts
  - **Develop** - New responses you need to learn for civilian success

4. Civilian Adaptation Strategy:

- For each "keep and adapt" response, write down how you could modify it for civilian contexts

- For each "replace" response, identify a healthier alternative
- For each "develop" response, create a plan for learning and practicing the new skill
- Specifically address how you'll handle situations where your military experience isn't automatically valued

5. Leadership Adaptation Plan:

- Identify situations where you need to lead or influence civilians
- Develop strategies for earning authority through competence and relationships rather than rank
- Practice empathetic communication and collaborative decision-making

6. Implementation Plan:

- Choose one stress response to work on first
- Create specific, achievable goals for practicing new responses
- Identify support resources (therapy, veteran groups, mentors, etc.)
- Set a timeline for evaluating progress

7. Support System Integration:

- Share your plan with trusted friends, family, or fellow veterans who can provide accountability
- Consider joining veteran transition groups or working with a counselor familiar with military transition

8. Regular Evaluation:

- After two weeks, assess which strategies are working and which need adjustment
- Continue to refine your approach based on what works best in your civilian environment

Self-Reflective Questions:

1. What stress responses served me well in the military but may be causing problems in civilian life?

2. How do I react when civilians don't automatically recognize or respect my military experience and background?

3. When I'm stressed, do I tend to isolate or connect with others? What might work better in civilian contexts?

4. How does the fullness of my Emotional Bucket affect my ability to respond appropriately to civilian stressors?

5. What civilian stressors do I find most challenging, and what military skills might help me address them more effectively?

6. How can I transform my military leadership experience into civilian influence and authority?

Affirmations:

1. I am capable of adapting my military strengths to civilian success.
2. Each challenge is an opportunity to practice healthier stress responses.
3. I have the skills and resilience to build a stress baseline that serves my civilian goals.
4. I can earn respect and authority in civilian contexts through my actions and character.
5. My military experience is valuable, even when others don't immediately recognize it.

Actionable Steps:

1. **Daily Stress Check-ins:** Take a moment each day to assess your stress level and consciously choose your response rather than reacting automatically.
2. **Practice Civilian Communication:** Work on adapting your communication style from command-oriented to collaborative, especially in stressful situations.
3. **Build Civilian Support Networks:** Actively cultivate relationships with civilians who can provide different perspectives on stress management.
4. **Develop Emotional Processing Skills:** Practice identifying and expressing emotions rather than automatically suppressing them.
5. **Maintain Physical Fitness:** Continue using exercise as a stress management tool but adapt your approach to civilian fitness goals rather than military requirements.
6. **Practice Earning Authority:** Focus on building influence through competence, relationships, and empathy rather than expecting automatic respect.

Implementation Strategies:

Weeks 1-2: Assessment

- Track current stress responses without trying to change them
- Identify patterns and triggers, especially related to authority and recognition
- Assess which responses are helping or hindering civilian goals

Weeks 3-4: Initial Adaptation

- Begin practicing one new stress response
- Modify one military response for a civilian context
- Practice leading or influencing someone using civilian-appropriate methods
- Seeking feedback from trusted civilian friends or family

Weeks 5-8: Integration

- Expand to practicing multiple new responses
- Continue refining approaches based on effectiveness
- Build consistency in healthier stress management
- Focus on building civilian authority through competence and relationships

Ongoing: Continuous Improvement

- Regularly assess and adjust approaches
- Seek additional training or support as needed
- Share your learning with other veterans in transition
- Continue developing civilian leadership and influence skills

By understanding and adapting your stress baseline, you can maintain the strengths developed through military service while building new capabilities for civilian success. The goal is not to abandon what made you effective in the military, but to expand your toolkit for the different challenges and opportunities of civilian life, including the reality that civilian authority must be earned rather than automatically granted.

Evidence-Based Research: Studies on veteran transition show that those who successfully adapt their military stress management skills to civilian contexts report better adjustment outcomes, improved relationships, and greater career satisfaction (Sayer et al., 2010). Research also indicates that veterans who learn to recognize and modify their stress responses show reduced symptoms of anxiety and depression during transition (Pietrzak et al., 2011). Additionally, studies on veteran workplace integration demonstrate that those who adapt their leadership style to civilian expectations achieve higher levels of influence and career advancement (Keeling et al., 2018).

The Power of Unconditional Love and Self-Care

Chapter

4

Emotional wellness during your transition from military to civilian life is an ongoing journey that demands consistent nurturing and attention. This chapter explores how unconditional love and dedicated self-care form the bedrock of emotional stability and personal joy for veterans. By integrating these practices into your daily civilian routine, you foster a resilient foundation that supports lasting emotional health during and after your transition.

Unconditional Love: The Spiritual Heart of Emotional Wellness for Veterans

Unconditional love, especially towards oneself, is fundamental to emotional well-being. Military service often requires compartmentalizing emotions and pushing through pain for mission success. While these skills served you well in service, civilian life offers the opportunity to practice self-compassion and acceptance. Unconditional love requires a complete acceptance of yourself, your service experiences, mistakes, growth, and all, without judgment. This profound acceptance facilitates navigating civilian transition with greater compassion and understanding, essential not just for yourself but also in relationships with civilian family, friends, and colleagues.

Types of Love in Veteran Context:

- **Storge (Familial Love):** Natural affection among family members, which may need rebuilding or strengthening after deployments and military separation. Example: "I know my son has made deplorable choices; however, I will never stop loving him."

- **Philia (Friendship):** Shared goodwill between friends, including both military friendships and new civilian relationships. Example: "I love you, Brother. Safe travels and let's keep in touch."

- **Agape (Universal Love):** Selfless love that is extended to all people, closely related to spirituality and service. Example: "I love you, Sister. You are a human being who is doing the best you can with what life has given you. You deserve unconditional love and kindness."

- **Eros (Romantic Love):** Passionate love that includes physical and emotional intimacy, which may need nurturing after military service challenges. Example: "I love you so much, sweetheart, and I want to build a life together."

Spiritual Practice and Unconditional Love for Veterans:

Many veterans find that loving unconditionally is a profoundly spiritual experience, especially when processing difficult service experiences. Military service often involves life-and-death decisions, moral complexities, and experiences that challenge one's faith or spiritual beliefs. Struggles in spiritual connections during transition usually reflect challenges in giving and receiving unconditional love, both from others and from yourself.

Spiritual practices can enhance your ability to love unconditionally, contributing to emotional healing and a higher personal frequency and vibration. This is particularly important when dealing with service-related guilt, survivor's guilt, or moral injury. Learning to love yourself unconditionally despite difficult service experiences is crucial for a healthy civilian transition.

Actionable Steps:

1. Reflect daily on your actions and thoughts during transition to foster a non-judgmental understanding of yourself as you adapt to civilian life.

2. Participate in veteran-friendly spiritual activities that enhance your sense of connectedness and belonging in both military and civilian communities.

3. Regularly engage in acts of kindness, seeing these as opportunities to practice agape while honoring your continued service to others.

Affirmations:

1. I embrace and radiate unconditional love every day, honoring my service and my civilian journey.

2. My spiritual practice deepens my capacity for love and helps me process my military experiences.

3. I am a conduit of love in all my relationships, drawing on the brotherhood I learned in service.

Self-Care: Cultivating Daily Joy and Resilience During Transition

Self-care is crucial for maintaining your emotional bucket during civilian transition and should be an integral part of your daily routine. For veterans, effective self-care involves activities that cater to your mental, physical, and emotional well-being while honoring your military experience and supporting your transition to civilian life. This prevents emotional overload and helps maintain a consistent mood during the challenging transition period.

Importance of Daily Self-Care for Veterans:

Self-care is something that most people only do twice a week. However, self-care should be practiced every morning to start your day, as it brings out your joy and improves your mood and overall well-being. Daily self-care practices are essential for maintaining a positive mood and staying in a joyful state during the often-challenging transition from military to civilian life. These activities should not be occasional, but rather embedded in your everyday life, to manage transition stress and foster long-term emotional wellness effectively.

Military service often taught you to put mission and others first, sometimes at the expense of personal care. While this selflessness was valuable in service, civilian life requires you to prioritize your own well-being so you can be effective in your new civilian mission, whether that's in your career, family, or community involvement.

Veteran-Specific Self-Care Considerations:

- **Honoring Your Service:** Self-care activities should acknowledge and honor your military experience rather than trying to forget it
- **Structure and Routine:** Many veterans benefit from structured self-care routines that provide the organization with value
- **Physical Readiness:** Maintaining physical fitness, not for military standards but for personal health and stress management
- **Mental Processing:** Including activities that help process service experiences and civilian transition challenges
- **Social Connection:** Balancing time with fellow veterans who understand your experience and civilians who support your transition

Daily Self-Care Activities for Veterans:

Morning Routine Options:

- **Physical activities:** Weightlifting, aerobics, running, yoga, biking, walking, dancing, or swimming
- **Spiritual practices:** Praying, meditating, reading spiritual texts, or listening to spiritual messages
- **Mindfulness activities:** Visualizing your future civilian self, breathing exercises, or mindful walking
- **Creative expression:** Reading, journaling, or engaging in art and creative expression
- **Sensory comfort:** Burning scented candles, listening to music, or enjoying a peaceful morning coffee

Nature-Based Self-Care:

- Walking on the beach or in nature
- Gardening (many veterans find peace in growing things after experiences of destruction)
- Outdoor activities that provide space and perspective
- Caring for pets or horses (animal therapy can be particularly beneficial for veterans)

Technology-Assisted Self-Care:

- Listening to motivational speeches or podcasts on YouTube
- Using meditation apps designed for veterans
- Virtual reality experiences that provide calm environments
- Online veteran support communities

Actionable Steps:

1. Develop a morning self-care routine that includes at least one activity from each category: physical, spiritual, and mindful practices.
2. Set reminders to take short self-care breaks throughout the day, especially during stressful civilian situations or when you notice your emotional bucket filling.
3. Dedicate time each week to engage in a longer self-care activity that brings you joy and helps process your transition experience.

Affirmations:

1. Daily self-care is my priority for maintaining joy and wellness during my civilian transition.
2. Each act of self-care enriches my life with peace and happiness, honoring my service.
3. I honor my body and mind with loving care each day, supporting my successful transition to civilian life.

Exercise
UNCONDITIONAL LOVE AND SELF-CARE FOR VETERANS

Objective: This exercise is designed to help veterans develop sustainable practices of unconditional love and self-care that support emotional wellness during civilian transition.

Instructions:

1. **Love Mapping for Veterans:** Create a "map" of your relationships, categorizing them under storge, philia, agape, and eros. Include both military and civilian relationships. Reflect on how each relationship contributes to or detracts from your emotional wellness during transition. Pay special attention to relationships that may need rebuilding or strengthening after military service.

2. **Military Experience and Self-Love Assessment:** Write a letter to yourself acknowledging your service, including both proud moments and difficult experiences. Practice unconditional self-acceptance by recognizing that all parts of your military experience contributed to who you are today.

3. **Daily Self-Care Audit:** For two weeks, track all your self-care activities each morning. Note which activities were most effective in boosting your mood and reducing stress during transitions. Identify patterns that work best for your individual needs.

4. **Veteran-Specific Self-Care Plan:** Design a personalized daily self-care routine that:

    - It takes 15-30 minutes each morning
    - Includes physical, spiritual, and mental components
    - Honor your military experience
    - Support your civilian transition goals
    - Can be maintained consistently

5. **Forgiveness and Love Integration:** Write a letter of forgiveness related to a military experience (you do not need to send it). Focus on expressing forgiveness fully to clear any harmful emotional residue that might be affecting your civilian relationships. Be sure to burn the letter in the kitchen sink after reading it three times.

6. **Daily Affirmations and Reflections:** Create personal affirmations that combine self-love with transition goals. Start and end each day with these affirmations. Reflect each evening on how these affirmations influenced your civilian interactions and self-perception throughout the day.

7. **Mindfulness Breaks:** Implement three 5-minute self-care breaks into your daily civilian routine (morning, midday, evening). Observe and note any changes in your stress level and emotional state during civilian activities.

8. **Community Connection:** Identify and connect with veteran-friendly spiritual or support communities that can reinforce your practice of unconditional love and self-care.

Weekly Reflection Questions:

- How has daily self-care affected my civilian transition experience?
- What forms of unconditional love am I finding most challenging to practice?
- How has my relationship with myself changed through this practice?
- What self-care activities bring me the most joy and peace?
- How can I better integrate military values with civilian self-care practices?

Assessment Criteria:

- Consistency in daily self-care practice
- Depth of reflection on unconditional love in the veteran context
- Integration of military experience with civilian self-care
- Evidence of improved emotional management during transition

Adopting a lifestyle centered around unconditional love and consistent self-care is key to sustainable emotional wellness during and after military transition. These practices not only help manage the emotional challenges of civilian adjustment but also enhance your overall quality of life, leading to a more fulfilled and balanced post-military existence. Remember, the goal is continuous improvement and adaptation, not perfection. Just as you trained consistently for military readiness, emotional wellness requires daily training for civilian readiness.

Self-Reflective Questions

1. **Introspection on Unconditional Love:** How do I demonstrate unconditional love towards myself and others daily during my civilian transition? Are there situations where maintaining this kind of love is challenging, particularly when dealing with civilian misunderstandings about military experience?

2. **Evaluation of Self-Care Practices:** What are my current self-care practices, and how effectively do they address my physical, mental, and emotional needs during this transition period?

3. **Impact of Love on Relationships:** How does my ability or inability to express unconditional love affect my relationships with family, friends, and new civilian acquaintances who may not understand military culture?

4. **Barriers to Self-Care:** What are the biggest obstacles that prevent me from engaging in regular self-care routines during my transition, and how can I overcome them?

5. **Link Between Spiritual Practices and Love:** How do my spiritual practices enhance my capacity to give and receive love during this challenging transition period? Are there spiritual activities I might explore to deepen this capacity while honoring my military experience?

Affirmations

1. **Embracing Love:** "I accept myself fully, including my military experiences, and offer the same unconditional love to others in my civilian life."

2. **Commitment to Self-Care:** "Every day, I commit to actions that nurture my body, mind, and spirit as I transition to civilian life."

3. **Joy in Everyday Life:** "I find joy in everyday civilian activities while honoring my service and cherishing each moment of connection with loved ones."

Actionable Steps

1. **Morning Self-Care Routine:** Establish a consistent morning self-care routine that takes 15-30 minutes and includes physical, spiritual, and mindful components that set a positive tone for your civilian day.

2. **Cultivate Mindful Relationships:** Actively practice empathy and understanding in your civilian interactions. This might include listening more intently to civilian perspectives, withholding judgment about different life experiences, and offering support without conditions.

3. **Integrate Physical Activity:** Incorporate at least 30 minutes of physical activity into your daily routine to enhance both physical and emotional health, adapting military fitness standards to civilian wellness goals.

4. **Establish a No-Technology Zone:** Dedicate an hour before bed to unwind without screens. Use this time for reading, meditation, or engaging in a relaxing hobby that soothes your mind and helps you process the day's experiences.

5. **Seek Professional Guidance:** If you find it challenging to implement self-care or struggle with emotional barriers during transition, consider seeking guidance from a therapist or counselor familiar with veteran transition issues to develop effective strategies tailored to your needs.

Evidence-Based Research: Regular engagement in self-care and spiritual practices has been shown to significantly reduce stress and improve emotional and psychological well-being, particularly among veteran populations facing transition challenges (Kabat-Zinn, 2003; Pargament, 1997). Studies specific to veteran populations demonstrate that daily self-care practices correlate with improved civilian adjustment outcomes and reduced symptoms of depression and anxiety (Brenner et al., 2008).

Warrior Mindset in Civilian Life

Chapter 5

This chapter is crucial in your journey towards civilian success: adapting your Warrior Mindset to civilian life. The mindset you developed during military service profoundly influences your ability to cope with civilian challenges and seize new opportunities. In this chapter, you will explore the nature of the warrior mindset, understand how it can be both an asset and a potential barrier in civilian contexts, and gain practical tools to adapt this powerful mindset for civilian growth, resilience, and positive change.

Understanding the Warrior Mindset

The warrior mindset refers to the mental framework developed through military training and service that emphasizes mission accomplishment, tactical thinking, resilience under pressure, and unwavering commitment to objectives. This mindset served you well in military contexts, where clear hierarchies, defined missions, and structured environments were the norm. However, civilian life operates under different principles and requires adapting this powerful mindset to new contexts.

Components of the Military Warrior Mindset:

- **Mission-first mentality** - Prioritizing objectives above personal comfort or convenience
- **Tactical thinking** - Breaking down complex problems into manageable components
- **Resilience under pressure** - Maintaining effectiveness during high-stress situations
- **Loyalty and commitment** - Dedication to team, mission, and values
- **Adaptability** - Adjusting tactics while maintaining strategic focus
- **Physical and mental toughness** - Pushing through discomfort to achieve goals
- **Leadership readiness** - Prepared to take charge when situations demand it

Warrior Mindset vs. Civilian Growth Mindset

While the warrior mindset shares some similarities with what civilian psychology calls a "growth mindset," there are important distinctions that veterans must understand for successful civilian transition.

Warrior Mindset Characteristics:

- Mission-focused with clear objectives
- Hierarchy-based decision making
- Rapid response to changing conditions

- Team-oriented with defined roles
- Success is measured by mission completion
- Failure analyzed for tactical improvement
- Leadership through authority and competence

Civilian Growth Mindset Characteristics:

- Process-focused with evolving objectives
- Collaborative decision making
- Thoughtful response with stakeholder consideration
- Individual-oriented with flexible roles
- Success is measured by learning and development
- Failure is viewed as a learning opportunity
- Influence through relationship building and persuasion

Adapting the Warrior Mindset for Civilian Success

The goal is not to abandon your warrior mindset but to adapt it for civilian environments where different approaches to problem-solving, relationship-building, and goal achievement are valued.

1. Mission-First to Purpose-Driven:

- Transform your mission-first mentality into purpose-driven civilian goals
- Apply the same dedication you had to military missions to civilian career and personal objectives
- Maintain the warrior's commitment while embracing civilian purpose exploration

2. Tactical Thinking to Strategic Planning:

- Use your tactical breakdown skills for civilian project management and problem-solving
- Apply military planning principles to civilian goal setting and execution
- Translate operational thinking into civilian strategic advantage

3. Command Leadership to Influential Leadership:

- Adapt your leadership capabilities from authority-based to influence-based
- Use your natural leadership readiness while learning civilian collaboration styles
- Transform command presence into civilian executive presence

4. Team Cohesion to Relationship Building:

- Apply lessons from military unit cohesion to civilian team building
- Use your understanding of loyalty and commitment in civilian professional relationships
- Translate military brotherhood into civilian networking and mentorship

5. Resilience to Adaptability:

- Maintain your warrior resilience while developing civilian emotional intelligence
- Use your mental toughness for civilian challenges while learning to process emotions appropriately
- Apply military adaptability to civilian career pivots and life changes

Challenges of the Warrior Mindset in Civilian Life

Understanding potential challenges helps you navigate civilian environments more effectively:

Intensity Mismatch:

- Civilian environments may not require or appreciate the intensity level natural to warriors
- Learning to "dial down" intensity while maintaining effectiveness
- Understanding when warrior-level intensity is appropriate versus when it might overwhelm civilians

Decision-Making Speed:

- Military training emphasizes rapid decision-making, while civilian contexts often require more deliberative processes
- Balancing your ability to make quick decisions with civilian needs for consultation and consensus-building
- Learning when to slow down and when to leverage your rapid decision-making abilities

Perfectionism vs. Progress:

- Military standards of excellence may translate to civilian perfectionism that impedes progress
- Learning to embrace civilian concepts of "good enough" and iterative improvement
- Maintaining high standards while accepting civilian timelines and resource constraints

Communication Style:

- Direct military communication may need softening for civilian sensitivities
- Learning to communicate with influence rather than authority
- Adapting briefing skills to civilian presentation and persuasion needs

Leveraging Warrior Mindset Strengths in Civilian Life

Your warrior mindset provides significant advantages in civilian contexts when properly channeled:

Professional Advantages:

- **Project management** - Military planning and execution skills translate directly
- **Crisis management** - Your ability to remain calm under pressure is highly valued
- **Leadership roles** - Natural leadership capabilities with civilian adaptation
- **Goal achievement** - Mission-focused approach ensures civilian objective completion
- **Team building** - Understanding of unit cohesion principles builds strong civilian teams

Personal Advantages:

- **Resilience** - Mental toughness helps navigate civilian life challenges
- **Commitment** - Dedication to objectives ensures follow-through on personal goals
- **Adaptability** - Military-trained flexibility aids civilian life transitions
- **Service orientation** - Continued desire to serve translates to community involvement and meaningful work
- **Physical readiness** - Maintained fitness provides energy and confidence for civilian pursuits

Incorporating the Emotional Bucket Concept

Understanding how your warrior mindset relates to your Emotional Bucket is crucial for civilian success. When your bucket is full of unprocessed military experiences, you're more likely to default to rigid warrior responses that may not serve you in civilian contexts. When your bucket is well-managed, you can thoughtfully choose when to engage your warrior mindset and when to adapt to civilian approaches.

Impact of a Full Emotional Bucket on Warrior Mindset:

- **Hypervigilance** - Maintaining warrior alertness in safe civilian environments
- **Overreaction** - Responding to civilian challenges with military-level intensity
- **Isolation** - Withdrawing when civilian approaches don't match military expectations
- **Frustration** - Becoming angry when the civilian pace or processes feel inefficient

Benefits of Managing Your Emotional Bucket:

- **Flexible Response** - Choosing appropriate intensity levels for civilian situations
- **Enhanced Relationships** - Building civilian connections while maintaining warrior values
- **Career Success** - Leveraging warrior strengths while adapting to civilian workplace cultures
- **Personal Growth** - Continuing to develop beyond military identity while honoring service experience

Exercise

WARRIOR MINDSET ADAPTATION ASSESSMENT

Objective: This exercise helps veterans identify their current warrior mindset characteristics and develop strategies for adapting them to civilian contexts.

Instructions:

1. **Warrior Mindset Inventory:**
   - List your strongest warrior mindset characteristics from military service
   - Identify specific examples of how these characteristics served you in military contexts
   - Note which characteristics you rely on most heavily in current civilian situations
2. **Civilian Context Analysis:**
   - Describe civilian situations where your warrior mindset has been an asset
   - Identify civilian situations where your warrior mindset may have been counterproductive
   - Analyze feedback you've received about your approach in civilian environments
3. **Adaptation Opportunities:**
   - For each warrior mindset strength, brainstorm how it could be adapted for civilian success
   - Identify civilian skills that would complement your warrior mindset
   - Create specific plans for developing these complementary skills
4. **Practical Application:**
   - Choose one civilian situation where you want to apply your adapted warrior mindset
   - Develop a specific plan for engaging this situation with both warrior strengths and civilian adaptations
   - Set measurable goals for this application

5. **Feedback Integration:**

    - Identify trusted civilian colleagues, friends, or family members who can provide feedback
    - Ask for specific feedback on how your military background affects your civilian interactions
    - Create a plan for integrating this feedback while maintaining your core warrior values

6. **Continuous Development:**

    - Establish a regular review process for assessing your warrior mindset adaptation
    - Identify resources (books, courses, mentors) that can help you continue developing civilian applications
    - Set goals for ongoing development that honor both your military background and civilian objectives

Self-Reflective Questions:

1. Which aspects of my warrior mindset serve me best in civilian life, and which need adaptation?

2. How do I react when a civilian approaches conflict with my military-trained instincts?

3. What feedback have I received about my intensity or approach in civilian settings?

4. How can I maintain my warrior values while adapting to civilian relationship and communication styles?

5. In what civilian contexts do I feel most confident applying my military-developed skills?

Affirmations:

1. I honor my warrior training while adapting skillfully to civilian contexts.
2. My military experience provides valuable strengths that I can apply thoughtfully in civilian life.
3. I am capable of maintaining my core values while growing in new directions.

Actionable Steps:

1. **Intensity Calibration:** Practice adjusting your intensity level based on civilian context. Start each civilian interaction by assessing the appropriate energy level for the situation.
2. **Communication Adaptation:** Work on softening your communication style for civilian contexts while maintaining the clarity and directness that served you in the military.
3. **Civilian Skill Development:** Identify and actively develop civilian skills that complement your warrior mindset, such as collaborative decision-making or relationship-building.
4. **Feedback Seeking:** Regularly ask trusted civilians for feedback on how your military background affects your interactions and use this feedback for continuous improvement.
5. **Value Integration:** Create a personal mission statement that integrates your warrior values with your civilian goals, providing guidance for decision-making in ambiguous situations.

Implementation Strategies:

Weeks 1-2: Assessment and Awareness

- Complete the warrior mindset inventory
- Observe your automatic responses in civilian situations
- Gather feedback from trusted civilian contacts

Weeks 3-4: Targeted Adaptation

- Choose one warrior mindset characteristic to adapt
- Practice modified approaches in low-stakes civilian situations

- Document what works and what needs further adjustment

Weeks 5-8: Integration and Expansion

- Expand adapted approaches to more significant civilian challenges
- Continue developing complementary civilian skills
- Build confidence in your ability to flexibly apply your warrior mindset

Ongoing: Mastery and Mentorship

- Continue refining your ability to adapt a warrior mindset to various civilian contexts
- Mentor other veterans in making similar adaptations
- Lead civilian teams using your adapted warrior leadership approach

Your warrior mindset is not something to be discarded in civilian life but rather a powerful tool to be thoughtfully adapted. By understanding both its strengths and its potential limitations in civilian contexts, you can leverage this mindset for continued success while building the additional skills needed for civilian achievement. The goal is to remain a warrior while becoming an effective civilian, maintaining your core strength while expanding your tactical repertoire for new environments.

Evidence-Based Research: Studies on veteran workplace integration show that those who successfully adapt their military mindset to civilian contexts while maintaining core values report higher job satisfaction and career advancement rates (Castro & Kintzle, 2014). Research also indicates that veterans who learn to flex their leadership and communication styles for civilian environments achieve better relationship outcomes and reduced conflict in personal and professional settings (Blackburn, 2016).

Notes

____

____

____

____

____

Finding Purpose after Service

Chapter

6

Purpose acts as the unseen force guiding our decisions and fueling our passions, infusing meaning and direction into our actions. For veterans, recognizing and embracing a new purpose after military service can transform the transition experience from merely finding civilian employment to embarking on a journey of continued personal and professional fulfillment. This chapter delves into the nature of post-military purpose, its pivotal role in crafting a fulfilling civilian life, and offers insights on how veterans can start to uncover and articulate their unique civilian purpose while honoring their military service.

Defining Purpose After Military Service

Purpose is the central, self-organizing life aim that shapes goals, provides direction, and fuels motivation beyond your military identity. It transcends mere career objectives to encompass all life facets, including personal growth, relationships, and continued service to others. Dynamic and evolving, post-military purpose is grounded in a deep awareness of the skills, values, and experiences gained through military service and how they can meet civilian world needs. Living purposefully after service is linked to numerous benefits such as enhanced mental health, greater life satisfaction, and an increased sense of continued contribution.

For veterans, purpose often involves finding ways to continue serving - whether through careers that help others, community involvement, veteran mentorship, or family leadership. The challenge is translating the clear sense of mission you had in the military into a meaningful civilian direction.

The Significance of Purpose for Veterans

The quest for purpose after military service is particularly crucial for veterans because military life provides such a clear and compelling sense of mission. When that mission ends, many veterans experience what researchers call "moral injury" - not from traumatic events, but from the sudden absence of clear purpose and meaning that military service provided.

For veterans, a clear sense of civilian purpose can:

- **Ease the transition** by providing direction and motivation during difficult adjustment periods
- **Strengthen civilian commitment** by connecting daily activities to meaningful goals
- **Inform career planning** by aligning civilian work with values developed during service
- **Enhance relationships** by providing common ground with civilian family and friends
- **Maintain service identity** by continuing to serve others in new contexts

Purpose aids in prioritizing efforts, ensuring that time and energy align with what truly matters while honoring the service values you developed in the military.

Benefits of Embracing Purpose After Service

- **Enhanced Motivation and Resilience**: Purpose serves as a wellspring of perseverance, enabling veterans to navigate civilian challenges with the same dedication they brought to military service.
- **Improved Mental and Physical Health**: Studies indicate that veterans with a strong sense of post-military purpose experience reduced depression and anxiety and report better physical health outcomes.
- **Increased Life Satisfaction**: Living in alignment with purpose that honors military experience while embracing civilian opportunities leads to deeper fulfillment and happiness.
- **Continued Service Impact**: Purpose allows veterans to continue making a positive difference, maintaining the service orientation that defined their military experience.

Types of Post-Military Purpose

Mission-Continuation Purposes:

- **Direct Service Roles**: Careers in law enforcement, firefighting, emergency services, or homeland security that continue protecting others
- **Veteran Support**: Working with veteran organizations, VA services, or veteran-owned businesses
- **Community Service**: Elected office, non-profit leadership, or community organization involvement
- **Education and Training**: Teaching, coaching, or training that develops others' capabilities

Skills-Translation Purposes:

- **Leadership Development**: Using military leadership experience to develop civilian leaders in corporate or community settings
- **Project Management**: Applying military planning and execution skills to civilian projects that improve communities or organizations
- **Crisis Management**: Using military crisis response training in civilian emergency management or business continuity roles
- **Technical Expertise**: Leveraging military technical training in civilian industries that advance technology or infrastructure

Values-Driven Purposes:

- **Family Leadership**: Focusing on being an exemplary parent, spouse, or family member as your primary mission
- **Mentorship**: Guiding younger veterans, at-risk youth, or developing professionals
- **Entrepreneurship**: Building businesses that embody military values while serving civilian needs
- **Advocacy**: Supporting causes that align with values developed during military service

Discovering Your Post-Military Purpose

Identifying your post-military purpose is a deeply personal journey involving introspection, experimentation, and continual refinement. Here are the steps to guide this exploration:

1. Reflect on Military Service Meaning:

- Identify what aspects of military service gave you the greatest sense of purpose and fulfillment
- Consider which military experiences made you feel most alive and engaged
- Analyze what you miss most about military service and how those elements might translate to civilian contexts

2. Clarify Core Values:

- Determine which values from military service remain most important to you
- Consider how these values might be expressed in civilian contexts
- Identify civilian environments that would honor and utilize these values

3. Assess Skills and Passion Intersection:

- Explore where your military-developed abilities and civilian interests meet
- Consider civilian needs that your unique combination of skills and experience could address
- Look for opportunities where your passion and competence align with civilian impact

4. Experiment and Engage:

- Try different civilian roles through volunteer work, internships, or informational interviews
- Engage with veteran service organizations to explore continued service opportunities
- Participate in civilian activities that might reveal new sources of meaning and purpose

5. Seek Veteran Mentorship:

- Connect with veterans who have successfully found meaningful civilian purposes
- Learn from their transition experiences and purpose-finding journeys
- Consider how their approaches might apply to your unique situation and interests

6. Set Purpose-Aligned Goals:

- Develop civilian goals that reflect your emerging sense of purpose
- Ensure these goals honor your military experience while embracing civilian opportunities
- Create action plans that move you systematically toward purpose-aligned activities

7. Regular Reflection and Reassessment:

- Continually reflect on your evolving sense of purpose as you grow in civilian life
- Adjust your direction based on new experiences and insights
- Remain open to purpose evolution while maintaining core values from military service

Incorporating the Emotional Bucket Analogy

Understanding how emotional management affects your journey toward finding post-military purpose is crucial. The Emotional Bucket analogy illustrates how emotional overload can obscure clarity and focus, essential for identifying your civilian purpose while processing military transition.

Impact of a Full Emotional Bucket:

- **Reduced Clarity and Focus**: Overwhelm from transition stress can cloud judgment, complicating the introspection needed for purpose identification
- **Obstructed Self-Discovery**: Emotional clutter from unprocessed military experiences prevents deep engagement with civilian possibilities
- **Diminished Motivation**: Excessive emotional load can sap energy, making the pursuit of purpose seem daunting when you're struggling with basic transition challenges

Benefits of Managing Your Emotional Bucket:

- **Enhanced Clarity**: Managing emotional load clears mental space, allowing for focused reflection on civilian aspirations and values that build on military experience
- **Openness to Exploration**: With emotional distractions minimized, exploring various civilian interests becomes more feasible, aiding in the discovery of what resonates deeply
- **Increased Energy and Motivation**: Reduced stress levels boost overall energy, invigorating your pursuit of purpose-driven civilian activities

This structured exploration and management of emotions facilitates a more straightforward pathway to discovering and living your post-military purpose, ultimately enhancing both personal satisfaction and successful civilian transition.

Exercise

FIND YOUR POST-MILITARY PURPOSE

Materials needed: Pen and paper

This exercise will help you discover your post-military purpose in as little as 15 minutes. What you discover as your purpose today may evolve as you continue to explore civilian life and discover new ways to apply your military skills and values to civilian contexts.

Career and Service Questions:

1. If you were guaranteed success in any civilian career field, which career field would you choose if someone paid you $500,000 per year to do it?

Note: If you know the answer to the question above, then you are very close to finding your post-military purpose.

2. Once you start working in your dream civilian career or building your dream business, how can you help/mentor other veterans who may have similar dreams?

Key Questions to Finding Your Post-Military Purpose:

(Sample answers provided in parentheses)

Who are you? *(Marine veteran, husband, and father)*

___

What do you enjoy doing, and what makes you come alive? *(Leading teams, solving complex problems, and mentoring others)*

___

Who do you enjoy doing it for? *(Young veterans transitioning to civilian life)*

___

What do the people you do it for want out of it? *(They want guidance on how to transition and find meaningful civilian careers successfully.)*

___

___

What's in it for you? *(I feel fulfilled continuing to serve and make a difference in people's lives)*

____

Sample Purpose Statement:

> *"Hello, my name is [Name], and I am a [branch] veteran who enjoys leading teams, solving complex problems, and mentoring others. I work primarily with young veterans transitioning to civilian life. These veterans want guidance on how to successfully transition and find meaningful civilian careers, and helping them makes me feel fulfilled because I'm continuing to serve and make a difference in people's lives."*

When asked, "So what do you do?" your response becomes: "I help young veterans successfully transition to meaningful civilian careers."

Your Purpose Statement:

____

____

____

____

The sooner you find your post-military purpose, the sooner you will find fulfillment in civilian life. Living a purpose-filled life that builds on your military experience will bring you lasting happiness and a continued sense of contribution.

You may need to take civilian employment to pay bills and support your family while ***working on your purpose*** part-time or as a volunteer. The goal is to find civilian work that aligns with your purpose (getting paid for what you love to do while helping people) and eventually building your own business or role where you're compensated for following your purpose. If financial security is established, volunteering to help change lives while living your purpose can be equally fulfilling.

Self-Reflective Questions

1. When do I feel the clearest and most focused about my civilian future? Can I link these moments to times when my Emotional Bucket feels lighter?

2. What civilian activities have I engaged in that felt most meaningful, and what was the state of my Emotional Bucket at these times?

3. Have there been moments when emotional overload from transition prevented me from pursuing interests that could be linked to my post-military purpose?

4. What strategies have been effective in managing my Emotional Bucket, and how might these assist in my journey to find my civilian purpose?

5. In what ways can I regularly empty my Emotional Bucket to better facilitate self-exploration and the discovery of my post-military purpose?

Affirmations

1. I have the clarity and focus to explore my true post-military purpose.
2. I commit to civilian actions that align with my deepest values and honor my military service.
3. I am capable of managing my emotions effectively to pursue my civilian calling while honoring my military experience.

Actionable Steps

1. **Practice Regular Mindfulness**: Engage in daily mindfulness practices to help maintain clarity by managing the contents of your Emotional Bucket during transition.
2. **Journal Your Emotional State**: Keep a daily journal of your emotions and civilian activities to identify correlations between a well-managed Emotional Bucket and moments of insight related to your purpose.
3. **Engage in Purposeful Exploration**: Schedule regular times to engage in new civilian activities that might align with your potential purposes, especially when your Emotional Bucket is well-managed.
4. **Develop Emotional Coping Skills**: Learn and practice emotional coping skills such as deep breathing, meditation, or speaking with a veteran-informed counselor to keep your Emotional Bucket from overflowing during transition.
5. **Reflect and Reassess**: Regularly reassess the contents of your Emotional Bucket and your progress in finding your post-military purpose to ensure alignment and adjust strategies as needed.

Evidence-Based Research Support

Research suggests that emotional clarity and management are critical in decision-making and self-reflection, which are essential when exploring post-military purpose. A study by Gross (2002) on emotion regulation highlights how managing one's emotional state can lead to better psychological health and enhanced life satisfaction, both crucial when finding and pursuing civilian purpose. Furthermore, research by Kashdan and McKnight (2009) demonstrates that veterans who find meaningful post-military purposes that align with their service values report significantly better transition outcomes and reduced symptoms of depression and anxiety.

Uncovering your post-military purpose is one of the most enriching endeavors you can undertake as a veteran, offering not only personal enrichment but also the capacity to continue positively impacting the world. Though the process can be complex, the clarity and satisfaction derived from discovering your civilian purpose while honoring your military service are immensely rewarding. Continue to refine and understand what drives you as you progress through civilian transition and into your new mission.

Notes

Elevating Your Frequency and Vibration

Chapter

7

The Law of Frequency and Vibration illustrates foundational principles in physics and metaphysics that can significantly enhance your post-military life. It posits that everything in the universe, from the minutest particles to the largest entities, vibrates at specific frequencies. Understanding and applying these principles can significantly enhance your civilian transition, personal relationships, and professional success by influencing your mental, emotional, and physical states.

Understanding Vibrational Frequency for Veterans

Vibrational frequency, as understood in metaphysics and holistic wellness, describes how fast the particles of an object or being vibrate. Rooted in the scientific principle that everything is composed of energy, this concept extends to suggest that this energy manifests in various forms and frequencies, affecting everything from physical objects to thoughts and feelings. Higher frequencies correlate with states of joy, peace, and enlightenment, while lower frequencies are associated with negative emotions like fear, anger, and grief.

For veterans, understanding frequency and vibration becomes particularly relevant during civilian transition. Military service often required operating at high-stress frequencies that, while necessary for mission success, may not serve you well in civilian contexts. Learning to consciously adjust your vibrational frequency can help you navigate civilian relationships, workplace dynamics, and personal challenges more effectively.

The Law of Frequency and Vibration

This law asserts that every element in the universe vibrates at a particular frequency and that similar frequencies attract each other. It underlines the potential of influencing your reality by altering the frequencies of your thoughts, emotions, and actions. For veterans transitioning to civilian life, mastering this law can be a powerful strategy to navigate the challenges of leaving military structure, finding new purpose, and building civilian relationships.

Why This Law Matters for Veterans

Applying the Law of Frequency and Vibration empowers veterans to proactively influence their civilian environments and outcomes. During military service, you operated within established structures and clear missions. In civilian life, this law provides a framework for creating positive experiences and attracting opportunities that align with your post-military goals.

Benefits for Veterans

1. **Enhanced Emotional Regulation**: Adjusting vibrational frequencies helps veterans manage the emotional challenges of transition more effectively.

2. **Improved Civilian Integration**: High vibrational frequencies improve social interactions and help veterans connect more naturally with civilian communities.

3. **Better Professional Relationships**: Higher frequencies attract positive people and foster harmonious relationships, crucial in civilian workplace environments.
4. **Manifestation of Transition Goals**: This law facilitates the realization of civilian aspirations by aligning thoughts and actions with desired post-military outcomes.
5. **Reduced Transition Stress**: Operating at higher frequencies naturally reduces anxiety and stress associated with major life transitions.

Military vs. Civilian Frequency Considerations

Military Frequency Patterns:

- High-stress operational frequencies designed for crisis response
- Intense focus frequencies for mission-critical situations
- Group cohesion frequencies that bond military units
- Alert frequencies for threat detection and response
- Discipline frequencies for following orders and protocols

Civilian Frequency Patterns:

- Collaborative frequencies for teamwork and consensus building
- Creative frequencies for innovation and problem-solving
- Social frequencies for relationship building and networking
- Peaceful frequencies for work-life balance and personal fulfillment
- Adaptive frequencies for navigating ambiguous situations

Mastering the Law of Frequency and Vibration as a Veteran

1. **Spiritual Practices for Veterans**: Engaging in spirituality and prayer connects you with a higher power, offering guidance and peace during transition. This provides control over your internal state, regardless of external civilian challenges. Many veterans find that spiritual practices help them process service experiences while elevating their frequency for civilian success.

2. **Mindfulness and Veteran-Specific Awareness**: Developing keen awareness of your mental and emotional states through mindfulness practices helps you remain present and engaged with your current civilian reality rather than being pulled back into military mindsets that may not serve your current situation.

3. **Intentional Thought and Emotion Management**: Actively choosing positive thoughts and emotions helps maintain a high vibrational state, enhancing overall well-being during transition. This involves consciously releasing negative thought patterns developed during high-stress military service.

4. **Civilian Environment Influence**: Surrounding yourself with positive civilian influences such as motivational content, supportive veteran communities, and uplifting civilian relationships can significantly boost your frequency and ease transition challenges.

5. **Physical Health for Veterans**: Regular exercise adapted to civilian fitness goals and proper nutrition play crucial roles in maintaining high energy levels and mental clarity during transition.

6. **Regular Practice**: Consistency in applying these techniques transforms them into habits that support ongoing civilian success and well-being.

7. **Veteran-Specific Self-Care**: Engaging in activities that bring joy while honoring your military experience enhances your mood and supports your vibrational frequency, contributing to successful civilian integration.

Incorporating the Emotional Bucket Analogy

Understanding how emotional overload can impact vibrational frequency is crucial for veterans. A "full bucket" of negative emotions from military service or transition challenges can lower your vibrational state, leading to negative civilian experiences. Conversely, managing emotional load effectively can maintain a higher, more positive frequency, attracting better civilian opportunities and relationships.

Impact of a Full Emotional Bucket on Frequency:

- **Lowered Vibration**: Unprocessed military trauma or transition stress creates heavy, low-frequency emotional states

- **Negative Attraction**: Lower frequencies tend to attract challenging situations and difficult people

- **Civilian Integration Challenges**: Low frequencies can make civilian environments feel foreign or unwelcoming

- **Professional Obstacles**: Lower vibrational states may repel positive career opportunities

Benefits of Managing Your Emotional Bucket:

- **Elevated Frequency**: Processing and releasing emotional baggage raises your natural vibrational state

- **Positive Magnetism**: Higher frequencies attract supportive people and beneficial opportunities

- **Enhanced Civilian Connection**: Elevated vibrations help you resonate with positive civilian communities

- **Career Success**: Higher frequency states attract mentors, opportunities, and professional growth

Practical Applications for Veterans

Goal Setting and Vision Boards: Visualizing desired civilian outcomes helps maintain focus and keeps your vibrational frequency aligned with your post-military aspirations. Create vision boards that honor your military service while depicting your civilian goals.

Visualization and Meditation: Regular practice aligns your mental state with your civilian aspirations, enhancing your ability to manifest transition goals. Many veterans benefit from guided meditations specifically designed for military transition.

Gratitude Practices for Veterans: Focusing on gratitude for both your military service and civilian opportunities enhances your vibrational state and attracts more positivity. This includes appreciation for military experiences while remaining open to civilian possibilities.

Veteran Community Accountability: Collaborating with other veterans to maintain high-frequency activities supports a positive environment conducive to successful transition and continued growth.

Exercise

LAW OF FREQUENCY FOR VETERANS

1. Which no-cost or low-cost self-care activities below lift your mood, spirit, or bring you joy while you served or during your civilian transition?

Physical Activities:

	Running or Jogging		Yoga
	Weightlifting		Sports
	Aerobics		Swimming
	Martial Arts		

Nature and Outdoor Activities:

	Walking in the Neighborhood		Gardening
	Walk the Beach		Connect with Nature
	Hiking		

Spiritual and Mindful Activities:

	Pray		Read Spiritual Books
	Meditate		Listening to Spiritual Messages
	Visualization		Listening to Spiritual Music

Creative and Relaxing Activities:

	Painting/Art		Scented Candles
	Music (listening or playing)		Dancing
	Reading books		

Social and Service Activities:

	Volunteer Work		Family Time
	Mentor Other Veterans		Community Service

Animal Care:

	Care for House Pets		Walk the Dog
	Care for Horses		

Other: ________________________________

2. Can these activities be done in less than 1 hour?
3. Can these activities be done every day as part of your morning routine?
4. Can you forgive people from your military service who hurt you or made poor decisions that affected you?
5. Can you set a date to work on forgiveness? (Example: June 1st)

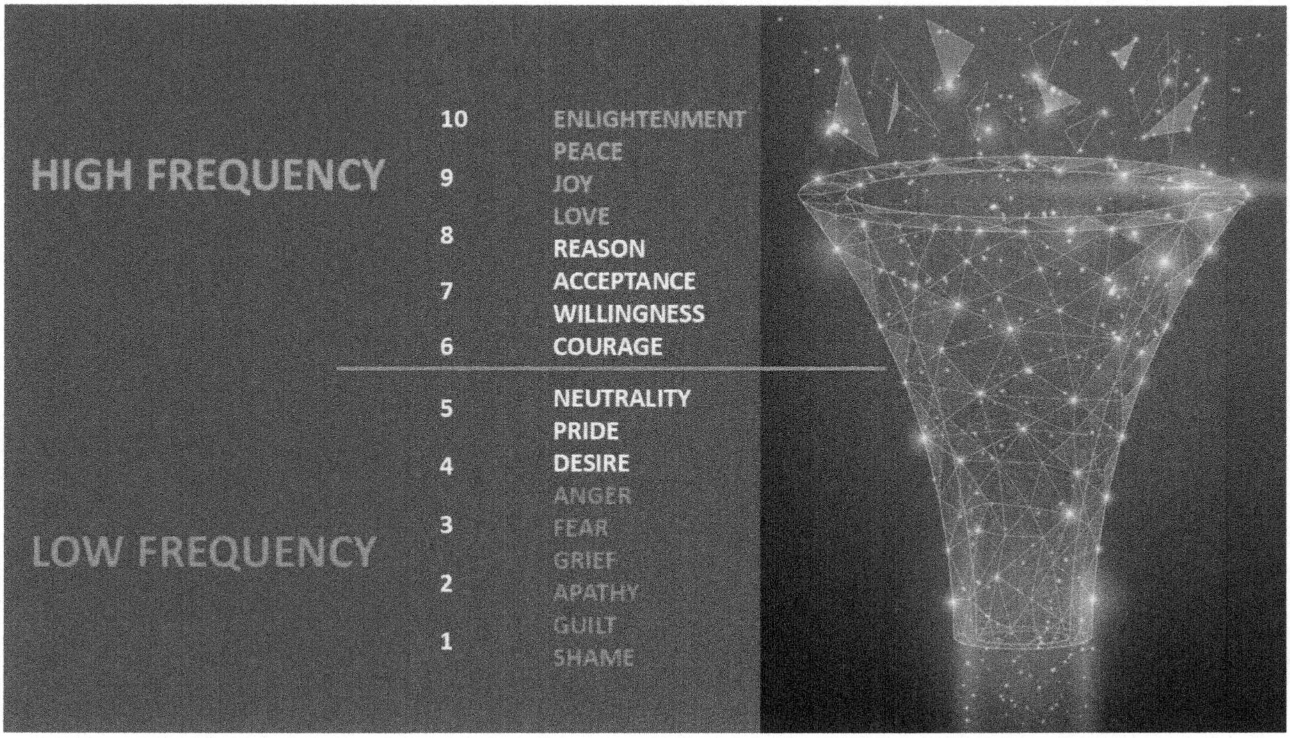

Frequency Elevation Plan:

Based on your selections above, create a daily routine that includes:

- One physical activity that elevates your frequency
- One spiritual or mindful practice
- One activity that brings you joy
- One practice that connects you with others in positive ways

Sample Daily High-Frequency Routine for Veterans:

- **Morning**: 20 minutes of meditation or prayer
- **Midday**: 30 minutes of physical activity
- **Evening**: Gratitude journaling focusing on both military service and civilian opportunities
- **Weekly**: Volunteer work or mentor other veterans

Self-Reflective Questions

1. Can I identify moments when a full Emotional Bucket might have lowered my vibrational frequency during civilian transition?

2. How do changes in my emotional state affect my daily civilian experiences and the people I attract?

3. What practices have I found effective in managing my Emotional Bucket and thus maintaining a higher frequency during transition?

4. In what ways do I notice the quality of my thoughts and civilian experiences shifting as I manage my emotions better?

5. What are some immediate actions I can take when I feel my vibrational frequency dropping during civilian challenges?

Affirmations

1. I am in control of my vibrational frequency through my thoughts and emotions during civilian transition.
2. I choose to fill my mind with positive thoughts that elevate my frequency and support my civilian success.
3. I attract positive energy and opportunities by maintaining a high vibrational state that honors both my military service and civilian potential.

Actionable Steps

1. **Practice Regular Mindfulness**: Engage in daily mindfulness or meditation to monitor and regulate your emotional state, helping to keep your Emotional Bucket and vibrational frequency balanced during transition.
2. **Adopt Positive Thought Patterns**: Consciously replace negative transition thoughts with positive affirmations and optimistic perspectives about your civilian future to maintain a higher vibration.
3. **Utilize Emotional Release Techniques**: Incorporate techniques like journaling, veteran-focused therapy, or emotional freedom technique (EFT) to manage and release emotions that lower your frequency.
4. **Cultivate Joyful Experiences**: Regularly engage in activities that bring you joy while honoring your military experience, as these help maintain a high vibrational frequency during transition.
5. **Connect with Positive Influences**: Spend time with people who uplift you, including both supportive veterans and positive civilian influences who share your desire to maintain an optimistic outlook and high vibrational state.

Evidence-Based Research Support

Research in the fields of psychology and neuroscience has shown that emotions can affect our physical reality and success outcomes. Studies like those conducted by the HeartMath Institute illustrate how positive emotions can enhance mental, emotional, and physical well-being, potentially aligning with the concept of vibrational frequencies affecting one's life experience (McCraty & Childre, 2010). Additionally, practices such as mindfulness and meditation have been scientifically proven to reduce stress and promote a positive mental state (Kabat-Zinn, 1994), supporting the idea that managing one's emotional state can influence their vibrational frequency and overall life outcomes.

For veterans specifically, research has shown that those who maintain positive emotional states and engage in regular spiritual or mindfulness practices report better transition outcomes, improved civilian relationships, and greater overall life satisfaction (Pietrzak et al., 2010). This supports the practical application of frequency and vibration principles in veteran civilian transition success.

Watch these life-changing videos to prepare you for your journey in understanding the Law of Frequency, Vibration, and Attraction:

Watch First

The Secret

Beyond the Secret

Beyond the Secret: The Moses Code

Beyond the Secret: The Awakening

SECTION II

Financial Wellness and Relationships

Mastering Post-Service Finances

Chapter

8

With special insights from Financial Advisor, Dexter Wyckoff

Financial management is a crucial skill that becomes even more critical for veterans transitioning from military to civilian life. The structured financial environment of military service, with its guaranteed pay, benefits, and clear advancement system, often leaves veterans unprepared for the complex financial decisions and uncertainties of civilian life. This chapter explores why developing healthy money habits is essential for veterans, outlines their benefits, and provides practical advice on mastering essential financial skills during and after military transition.

As financial advisor Dexter Wyckoff notes, "You've dedicated a portion of your adulthood to serving our country. You and your family have made remarkable sacrifices. Now, you face what will be the most daunting task of your adult life: assimilating into the civilian world. All of your training, experience, and combat time will feel like Basic Training when compared to what lies ahead. Here's the good news: there is still a path towards financial freedom."

Why Healthy Money Habits Are Critical for Veterans

For many veterans, civilian life represents the first experience of managing finances without the structure and predictability of military pay systems. Military service provides regular paychecks, guaranteed healthcare, housing allowances, and clear promotion-based pay increases. Civilian employment often lacks these certainties, requiring veterans to navigate variable income, employer-based benefits, and complex financial decisions that were previously handled by military systems.

Whether you satisfied one contract or served until retirement, your military service grants you access to unique financial benefits. It is important to understand how it all works, and developing healthy money habits helps veterans avoid common transition pitfalls such as excessive debt, poor credit, and financial stress that can detract from successful civilian integration and overall well-being.

Building a Foundation: Beliefs, Standards, and Decisions

Before we can understand how investments, insurance, and other financial products function, we must establish a healthy set of beliefs around money, saving, and investing, according to Dexter Wyckoff, having beliefs is important because they form the foundation of your standards, which will drive your decision-making process.

Beliefs → Standards → Decisions

Consider adopting these fundamental financial beliefs for veterans:

1. I believe that EVERYONE deserves to retire
2. I believe that systematic savings lead to guilt-free spending
3. I believe that 80% of our financial success is driven by behavior (NOT by income)
4. I believe that everyone should dedicate 20% of their net income to their financial plan
5. I believe that there is a future version of us that depends on the "Us" of today to take care of them
6. I believe that maximizing VA benefits is a critical component of financial success

These beliefs drive the standards that filter all financial decisions. Feel free to adopt them as your own and add beliefs that reflect your unique military experience and civilian goals.

The 20-60-20 Financial Framework for Veterans

Dexter Wyckoff recommends dividing veteran income into three strategic categories:

Financial Planning = 20%

	Retirement savings		Accelerated debt payments
	Education planning for dependents		Mid-and-term investments
	Building an emergency fund (3-6 months of your non-negotiable expenses)		

Non-Negotiable Expenses = 60%

	Food		Transportation
	Insurance		Utilities
	Basic clothing and necessities		Housing - (Rent/Mortgage should generally be 30% total net income)

Discretionary Expenses = 20%

	"Fun Money"		Non-essential purchases
	Entertainment		Dining out
	Hobbies		Travel and recreation

This framework provides structure similar to military planning while allowing flexibility for civilian lifestyle choices.

A Veteran Financial Success Story: Learning from Jimmy

Let's examine the financial journey of Jimmy, a Marine veteran who exemplifies successful post-service financial planning. Jimmy served 4 years as a Marine, now works for a start-up company, and is in his late 20s. He's been out of the military for a couple of years and finally feels like he has his footing.

Jimmy's financial profile includes:

- $60,000 annual salary from civilian employment
- $13,224 annual (50% VA disability) compensation (tax-free)
- Goals to purchase a home, start a family, and save for retirement
- Previous limited savings during military service
- No TSP contributions while serving

Understanding Veteran Tax Advantages

It's crucial to understand how different types of veteran income are taxed. Jimmy's earned civilian income is taxed at applicable state and federal rates, but his VA disability compensation is completely tax-free. This tax-free income creates unique opportunities for strategic financial planning.

Since Jimmy has earned income, he's eligible to contribute to a Roth IRA. Dollars contributed to a Roth IRA grow and distribute on a tax-free basis (following IRS guidelines). This means Jimmy can take his tax-free VA disability income and invest it in a Roth IRA, creating a powerful tax-free growth strategy.

Veteran-Specific Financial Opportunities

VA Home Loan Benefits: Jimmy's goal to purchase a home is supported by his access to the VA loan program. While he still needs to qualify for a mortgage (debt-to-income ratio, credit score, etc.), he won't be required to make a down payment. Since the loan is backed by the federal government, Private Mortgage Insurance (PMI) is not required. However, Jimmy still plans to put money down to keep his mortgage below 30% of his net monthly income, allocating part of his 20% financial planning budget toward this down payment.

Three-Pillar Retirement Strategy for Veterans

As Jimmy plans for retirement, Dexter Wyckoff emphasizes that there are three sources of income that will ensure veterans make it to and through retirement:

1. Base Incomes (45-50% of retirement income goal): Guaranteed and predictable sources of income not tied to stock or bond market performance:

- VA disability income (continues for life)
- Social Security benefits
- Military pensions (for retirees)
- Annuity income
- Other guaranteed income sources

2. Well-Diversified Investment Portfolio (50-55% of retirement income goal): A long-term investment strategy where after-tax returns outpace inflation. Jimmy must understand appropriate risk levels based on his time horizon. Historical data show the S&P 500 has never finished lower over any 15-year period than where it started, meaning long-term investors can expect fairly predictable growth rates.

3. Cash Reserve (3 years of projected retirement expenses): While long-term investing generally succeeds, markets experience periodic downturns. Having a cash reserve replaces investment portfolio income during unfavorable market conditions. Assuming retirement at 65 with 20-25 additional years of life expectancy, veterans should prepare to weather approximately three market crashes during retirement.

Veteran Life Insurance Considerations

As Jimmy plans to grow his family, life insurance becomes critical. He previously had Servicemember's Group Life Insurance (SGLI) during his Marine service, but civilian life requires different coverage.

If Jimmy were to pass away, his loved ones would need to:

- Pay off all consumer debt
- Replace his income until his partner retires
- Purchase a home or pay off existing mortgage
- Cover education costs for dependents

Life insurance policy claims are paid tax-free, allowing Jimmy's beneficiaries to maintain their lifestyle with minimal financial disruption. Importantly, Jimmy's VA disability income will end when he passes away, so his life insurance planning must replace both earned income and VA disability compensation.

Unique Financial Challenges for Veterans

Transition-Specific Challenges:

- Income variability from the military's guaranteed pay to civilian employment uncertainty
- Benefit from complexity navigating civilian healthcare, retirement plans, and insurance options
- Credit establishment building civilian credit history while maintaining military-established credit (Many times, being active military can give landlords comfort that you will pay, even with lower credit scores.)
- Geographic relocation costs associated with moving to civilian employment locations
- Career exploration, potential income reduction while exploring civilian career options

Post-Service Financial Opportunities:

- **VA benefits:** Maximizing educational, healthcare, and disability compensation benefits
- **Military skill premiums:** Leveraging military experience for higher civilian compensation
- **Security clearance value:** Monetizing security clearances in civilian employment
- **Veteran business opportunities:** Accessing veteran-specific entrepreneurship resources
- **Networking advantages:** Utilizing military networks for civilian career advancement

Benefits of Healthy Money Habits for Veterans

1. **Reduced Transition Stress**: Effective money management significantly reduces anxiety related to civilian financial uncertainty, allowing veterans to focus more on career development and personal relationships
2. **Maximized Veteran Benefits**: Understanding financial planning helps veterans fully utilize VA benefits, education benefits, and other veteran-specific financial resources

3. **Career Flexibility**: Strong financial habits provide the security to pursue meaningful civilian careers rather than just accepting any available employment

4. **Family Stability**: Financial competence supports family stability during the challenging transition period and beyond

5. **Entrepreneurship Options**: Sound financial management creates opportunities for veteran entrepreneurship and business ownership

Incorporating the Emotional Bucket Concept

Understanding how emotional management affects financial decision-making is crucial for veterans. Financial stress, whether from transition uncertainty or poor money management, can quickly fill your Emotional Bucket, leading to poor financial decisions, relationship strain, and reduced career performance.

Impact of a Full Emotional Bucket on Financial Habits:

- **Impulsive Spending**: Emotional distress from transition can lead to "retail therapy" or impulsive purchases for emotional relief

- **Avoidance Behaviors**: Financial overwhelm can lead to avoiding important financial decisions or planning

- **Risk Aversion**: Excessive emotional load may cause overly conservative financial decisions that limit growth

- **Relationship Strain**: Financial stress combined with transition stress can damage family and personal relationships

Benefits of Managing Your Emotional Bucket for Financial Success:

- **Improved Decision-Making**: A clear emotional state enhances the ability to make rational financial decisions

- **Increased Financial Discipline**: Reduced emotional stress makes it easier to stick to budgets and financial plans

- **Enhanced Long-Term Planning**: Managed emotions allow for better focus on long-term financial goals

- **Better Financial Communication**: Emotional wellness improves the ability to discuss finances with family and financial advisors

Exercise

VETERAN FINANCIAL PLANNING USING THE 20-60-20 METHOD

Objective: To develop a comprehensive financial plan using Dexter Wyckoff's 20-60-20 framework that addresses veteran-specific financial challenges and opportunities.

1. Apply the 20-60-20 Framework to Your Income

Monthly Income Analysis:

- Military/VA Income: $__________
- Civilian Income: $__________
- Total Monthly Income: $__________

20-60-20 Allocation:

- Financial Planning (20%): $__________
- Non-Negotiable Expenses (60%): $__________
- Discretionary Expenses (20%): $__________

2. Financial Planning Category Breakdown (Your 20%)

Priority Area	Monthly Allocation	Annual Goal
Emergency Fund	$__________	$__________
Retirement Savings	$__________	$__________
Home Down Payment	$__________	$__________
Debt Acceleration	$__________	$__________
Education Fund	$__________	$__________
Total	$__________	$__________

3. Three-Pillar Retirement Analysis

Base Incomes (Target: 45-50% of retirement needs):

- VA Disability: $___________monthly
- Future Social Security: $___________monthly (estimated)
- Military Pension: $___________monthly (if applicable)
- Other Guaranteed Income: $___________monthly
- **Total Base Income**: $___________monthly

Investment Portfolio Planning:

- Current Retirement Savings: $___________
- Monthly Contribution Goal: $___________
- Target Portfolio Value at Retirement: $___________

Cash Reserve Planning:

- Annual Retirement Expenses (projected): $___________
- 3-Year Cash Reserve Needed: $___________
- Current Progress Toward Cash Reserve: $___________

4. VA Benefits Maximization Checklist

Benefit Type	Eligible?	Applied?	Monthly Value	Status
Disability Compensation	Y/N	Y/N	$___________	___________
Educational Benefits	Y/N	Y/N	$___________	___________
Healthcare	Y/N	Y/N	$___________	___________
Home Loan Eligibility	Y/N	Y/N	$___________	___________

5. Life Insurance Needs Analysis

If you passed away tomorrow, your loved ones would need:

- Debt payoff: $__________
- Income replacement (annual): $__________x ______ years = $__________
- Home purchase/payoff: $__________
- Children's education: $__________
- **Total Life Insurance Need**: $__________

Current Coverage:

- Group Life Insurance: $__________
- Personal Life Insurance: $__________
- **Coverage Gap**: $__________

Self-Reflective Questions

1. How well does my current financial plan align with the 20-60-20 framework, and what adjustments do I need to make?

2. Am I maximizing all available VA benefits, and what steps do I need to take to ensure I'm not leaving money on the table?

3. How confident do I feel about my retirement planning using the three-pillar approach, and what areas need more attention?

4. What financial beliefs do I need to adopt or change to support my long-term financial success?

5. How can I better integrate my military experience and values into my civilian financial planning?

Affirmations

1. "I am capable of making disciplined financial decisions that honor my military service and support my civilian goals."
2. "My military training has prepared me to execute a strategic financial plan with precision and dedication."
3. "I am proactive about maximizing my veteran benefits and creating lasting financial security for my family."

Actionable Steps

1. **Implement the 20-60-20 Framework**: Immediately restructure your budget using Dexter Wyckoff's recommended allocation, ensuring 20% goes toward your financial future.
2. **Audit All VA Benefits**: Schedule appointments with VA representatives to ensure you're receiving all eligible benefits and understand how to maximize their value.
3. **Develop Your Three-Pillar Retirement Strategy**: Create specific plans for base income optimization, investment portfolio growth, and cash reserve accumulation.
4. **Seek Professional Financial Guidance**: Connect with financial advisors who understand veteran-specific financial planning and can help optimize your unique situation.
5. **Regular Financial Reviews**: Establish quarterly reviews of your financial progress, adjusting strategies based on changing circumstances and opportunities.

As Dexter Wyckoff emphasizes, "Remember, seek help from a licensed financial professional. It can be difficult to see your own picture while you are in the frame. ***Keep your family first*** and vision first and you will ALWAYS win."

By mastering post-service finances using proven frameworks and veteran-specific strategies, you can ensure that money management supports rather than hinders your civilian transition and long-term success. The goal is to build financial stability that provides security while creating opportunities for continued service and personal fulfillment in civilian life.

Evidence-Based Research: Research in behavioral economics shows that structured financial planning significantly improves transition outcomes for veterans (Smith et al., 2018). Studies also indicate that veterans who actively manage their financial transition report lower stress levels and greater satisfaction with civilian life (Johnson & Martinez, 2019). Additionally, research demonstrates that veterans who maximize their VA benefits and develop civilian financial literacy skills achieve better long-term financial stability and career success (Veterans Affairs Research Study, 2020).

Acknowledgment: Special thanks to Financial Advisor Dexter Wyckoff for his expertise and insights into veteran financial planning, including the 20-60-20 framework and three-pillar retirement strategy that form the foundation of this chapter's practical guidance.

Notes

VA Compensation: Maximizing Your Benefits

Chapter

9

Understanding and maximizing your VA compensation benefits is one of the most important financial decisions you'll make as a veteran. These benefits can provide substantial financial support throughout your lifetime, yet many veterans fail to receive the full compensation they've earned through their service. This chapter provides comprehensive guidance on navigating the VA compensation system, including the five essential steps to service connection, timing strategies, and how to build the strongest possible claim.

The Importance of VA Compensation

VA disability compensation is a tax-free monthly payment made to veterans who were injured or became ill because of their military service. Unlike many civilian benefits, VA compensation is not means-tested and continues for life, making it a crucial component of your long-term financial security. For many veterans, maximizing VA compensation can mean the difference between financial struggle and stability in civilian life.

Key Benefits of VA Compensation:

- **Tax-free income** for life
- **Automatic cost-of-living adjustments** to maintain purchasing power
- **Additional benefits** including healthcare, education, and vocational rehabilitation
- **Family benefits** for dependents in certain circumstances
- **Financial stability** during civilian transition and beyond

The Power of Starting Early: Two Years Before Separation

The most critical advice for active-duty service members is to begin your VA journey two full years prior to retirement or separation. This timeline allows you to document conditions, seek treatment, and build the medical evidence necessary for successful claims. Starting early is not about fabricating conditions - it's about ensuring that legitimate service-related health issues are properly documented and treated.

Why Two Years Matters:

- **Time to establish medical records** showing ongoing treatment and symptoms
- **Multiple medical appointments** to document the progression and impact of conditions
- **Specialist referrals** for complex conditions requiring expert evaluation

- **Treatment history** demonstrating attempts to address service-related health issues
- **Continuity of care** from active duty through veteran status

The Five Steps to Service Connection

Establishing service connection is the foundation of any successful VA disability claim. You must prove three essential elements, supported by two additional strategic steps:

Step 1: Current Disability or Condition You must have a current medical condition that can be diagnosed by a healthcare provider. This requires:

- **Medical diagnosis** from a qualified healthcare provider
- **Ongoing symptoms** that affect your daily life and work capacity
- **Documentation** through medical records, treatment notes, and specialist evaluations
- **Functional impact** showing how the condition limits your activities

Step 2: Service Connection or Nexus You must establish a connection between your current condition and your military service. This can be demonstrated through:

- **Direct service connection** for conditions that occurred during active duty
- **Secondary service connection** for conditions caused by other service-connected disabilities
- **Presumptive service connection** for certain conditions associated with specific exposures or service locations
- **Medical nexus opinion** from qualified healthcare providers linking the condition to service

Step 3: In-Service Event, Injury, or Illness You must show that something happened during your military service that caused or aggravated your current condition:

- **Service medical records** documenting injuries, illnesses, or treatments
- **Buddy statements** from fellow service members who witnessed events or symptoms
- **Service records** showing deployment locations, occupational exposures, or duties
- **Personal statements** describing when and how conditions began or worsened during service

Step 4: Comprehensive Medical Documentation Building a strong medical record is crucial for maximizing your benefits:

- **Regular medical appointments** for all conditions, not just major ones
- **Specialty care** when primary care providers recommend referrals
- **Complete symptom reporting** including seemingly minor issues that may worsen over time
- **Treatment compliance** following through with recommended therapies and medications

Step 5: Strategic Timing and Preparation Maximize your chances of success by timing your claim strategically:

- **Pre-separation filing** when possible to ensure continuity of benefits
- **Thorough preparation** of all evidence before submitting your claim
- **Professional assistance** from Veterans Service Organizations (VSOs) or accredited representatives
- **Appeal readiness** understanding the process if your initial claim is denied

The Critical Two-Year Strategy

Starting your VA journey two years before separation provides time to address health concerns systematically. This is not about seeking unnecessary medical care, but ensuring that legitimate service-related conditions are properly documented and treated.

What to Do During Your Final Two Years:

Medical Care Strategy:

- **Annual comprehensive physical exams** that address all health concerns
- **Document everything** from headaches and joint pain to sleep issues and stress
- **Seek treatment** for any condition affecting your health or performance
- **Don't minimize symptoms** when speaking with medical providers
- **Request specialist referrals** for ongoing or complex issues

Communication with Medical Providers: When speaking with military medical providers, be thorough and honest about your symptoms:

- **Describe all symptoms** you're experiencing, no matter how minor they seem
- **Explain functional impact** on your daily activities, sleep, work performance, and relationships
- **Request diagnostic tests** like X-rays, MRIs, CT scans, or laboratory work when appropriate
- **Ask for referrals** to specialists for ongoing issues
- **Follow up** on all recommended treatments and therapies

The Nurse Technician Strategy: One particularly effective approach is to communicate comprehensively with nurse technicians before the doctor enters the room. Nurse technicians often document symptoms in plain language in your medical records, which can be valuable evidence for future VA claims. They typically have more time to listen and may be more receptive to detailed symptom descriptions.

When speaking with nurse technicians, discuss:

- **All physical symptoms** from head to toe
- **Sleep disturbances** and their causes
- **Stress and mental health** concerns
- **Pain levels** and their impact on daily activities
- **Any changes** in your health since your last appointment

Common Conditions to Document:

Musculoskeletal Issues:

- **Joint pain** in knees, shoulders, back, ankles, wrists
- **Mobility limitations** affecting your ability to perform duties
- **Chronic pain** that persists despite treatment
- **Injury-related conditions** from training, deployment, or occupational duties

Mental Health Conditions:

- **Sleep disorders** including insomnia, nightmares, or sleep apnea
- **Stress and anxiety** related to service experiences
- **Depression** or mood changes since entering service (loss of a child, non-judicial punishment, divorce or alcoholism caused by too many deployments, etc.)
- **PTSD symptoms** from combat, or military sexual trauma
- **Adjustment difficulties** during deployments or transitions

Environmental Exposures:

- **Hearing loss or tinnitus** from noise exposure
- **Respiratory issues** from burn pits, chemicals, or other exposures
- **Skin conditions** from environmental or occupational exposures
- **Gastrointestinal problems** potentially related to deployment conditions

The PACT Act and Expanded Benefits

The Promise to Address Comprehensive Toxics (PACT) Act significantly expanded VA benefits for veterans exposed to burn pits and other toxic substances. This legislation creates new presumptive conditions for certain exposures, making it easier for affected veterans to receive compensation.

PACT Act Benefits:

- **Presumptive conditions** for burn pit exposure during post-9/11 deployments
- **Expanded Agent Orange presumptions** for Vietnam-era veterans
- **New covered locations** for environmental exposures
- **Streamlined claims process** for covered conditions

The Power of Secondary Conditions

Secondary conditions are medical conditions caused or aggravated by an already service-connected disability. Understanding and claiming secondary conditions can significantly increase your overall disability rating and compensation.

Common Secondary Condition Examples:

If you filed a claim and were awarded 10% for tinnitus, your next secondary claim could be for headaches caused by the ringing in your ears. Once you receive a 30% rating for migraines, you may then file a claim for depression resulting from the persistent migraine headaches. After being awarded 50% for depression, you could pursue a claim for sleep apnea, which may be linked to your depression and potential weight issues. This process can continue indefinitely. Additionally, medications prescribed for one condition can lead to secondary conditions such as headaches, erectile dysfunction, depression, irritable bowel syndrome (IBS), urinary incontinence, etc.

- **Sleep apnea** secondary to PTSD or depression
- **Depression** secondary to chronic pain conditions
- **Knee problems** secondary to back injuries
- **Diabetes** secondary to PTSD medications
- **Erectile dysfunction** secondary to diabetes or PTSD

Building Strong Secondary Claims:

- **Medical evidence** linking the secondary condition to the primary service-connected condition
- **Nexus letters** from healthcare providers explaining the relationship
- **Treatment records** for both primary and secondary conditions
- **Research support** showing known connections between conditions

Filing Strategies and Timelines

Minimum Service Requirements: You can file for VA compensation after serving only 91 consecutive days of active-duty service, provided you didn't receive a dishonorable discharge. This opens VA benefits to veterans with shorter service periods who may have sustained injuries or developed conditions during their brief service.

Optimal Filing Timeline:

- **180 days before separation:** Begin gathering medical evidence and preparing your claim
- **90 days before separation:** Submit your pre-discharge claim if possible

- **At separation:** Ensure all medical records are complete and transferred to VA
- **Post-separation:** File any additional claims for conditions that develop or worsen after service
- **Appeal Option:** If any of your claims are denied and you plan to fight the denial, ensure that you never allow the claim to expire, so refile/appeal (supplemental claim or higher-level review) within 12 months of a denial...avoid the Board of Veterans Appeals, the process could take years

Working with Veterans Service Organizations (VSOs)

VSOs provide free assistance with VA claims and have accredited representatives who understand the complex VA system. They can significantly improve your chances of success.

Benefits of VSO Assistance:

- **Expert knowledge** of VA regulations and procedures
- **Free services** with no cost to veterans
- **Advocacy support** throughout the claims and appeals process
- **Local representation** at VA regional offices
- **Appeals assistance** if your claim is initially denied

Common Mistakes to Avoid

Documentation Errors:

- **Waiting too long** to seek medical care for service-related conditions
- **Minimizing symptoms** when speaking with healthcare providers
- **Incomplete medical records** missing key appointments or treatments
- **Poor communication** with medical providers about functional impact

Filing Mistakes:

- **Incomplete applications** missing required forms or evidence
- **Missed deadlines** for filing appeals or additional evidence
- **Going it alone** without professional assistance from VSOs
- **Giving up too early** after initial denials rather than pursuing appeals

Exercise

VA COMPENSATION PLANNING

Objective: Create a comprehensive plan for maximizing your VA compensation benefits.

1. Service Timeline Assessment

Current Status: (Check one)

- ☐ Active duty with 2+ years remaining
- ☐ Active duty with less than 2 years remaining
- ☐ Recently separated (within 1 year)
- ☐ Veteran separated more than 1 year ago

Service History:

- Branch of service: ______________________
- Years of service: ______________________
- Deployment history: ______________________
- Occupational specialties: ______________________

2. Health Condition Inventory

Current Health Issues: (Document all conditions affecting you)

Body System	Condition/Symptom	Severity (1-10)	Service-Related?	Documented?
Musculoskeletal	______________	__________	Y/N	Y/N
Mental Health	______________	__________	Y/N	Y/N
Neurological	______________	__________	Y/N	Y/N
Cardiovascular	______________	__________	Y/N	Y/N
Respiratory	______________	__________	Y/N	Y/N
Gastrointestinal	______________	__________	Y/N	Y/N
Genitourinary	______________	__________	Y/N	Y/N

Body System	Condition/Symptom	Severity (1-10)	Service-Related?	Documented?
Skin	____________	____________	Y/N	Y/N
Other	____________	____________	Y/N	Y/N

3. Medical Documentation Plan

For each undocumented condition, plan:

- Medical appointment needed: ____________________
- Specialist referral required: ____________________
- Diagnostic tests needed: ____________________
- Timeline for documentation: ____________________

4. Service Connection Strategy

For each condition, identify:

- In-service event or exposure: ____________________
- Available service medical records: ____________________
- Buddy statements needed: ____________________
- Nexus evidence required: ____________________

5. Secondary Condition Analysis

Potential secondary conditions to research:

- Primary condition: ____________ → Secondary condition: ____________
- Primary condition: ____________ → Secondary condition: ____________
- Primary condition: ____________ → Secondary condition: ____________

Self-Reflective Questions

1. Have I been completely honest with medical providers about all my symptoms and their impact on my daily life?

2. What service-related health issues have I been ignoring or minimizing that should be documented?

3. How well do I understand the connection between my current health conditions and my military service?

4. What steps do I need to take to strengthen my medical evidence before filing VA claims?

5. Who can I connect with to get professional assistance with my VA claims process?

Affirmations

1. I deserve compensation for health conditions related to my military service.
2. I am proactive about documenting and treating my service-related health conditions.
3. I will advocate for myself to ensure I receive all the benefits I have earned through my service.

Actionable Steps

1. **Schedule Comprehensive Medical Exam**: Book an appointment focused on documenting all health concerns, no matter how minor they seem.
2. **Connect with a VSO**: Contact a Veterans Service Organization to get free professional assistance with your claims.
3. **Gather Service Records**: Collect all available service medical records, personnel records, and deployment documentation.
4. **Document Symptoms Daily**: Keep a journal of symptoms and their functional impact to provide detailed information to healthcare providers.
5. **Research Your Exposures**: Investigate potential exposures during your service that might be covered under the PACT Act or other presumptive conditions.

Key Takeaways

Maximizing your VA compensation requires strategic planning, thorough documentation, and persistence. The key is starting early, being completely honest about your health conditions, and seeking professional assistance when needed. Remember that VA compensation is not charity - it's compensation you've earned through your service and sacrifice.

Your VA compensation can provide financial security for life, but only if you take the necessary steps to document and claim the benefits you've earned. Don't leave money on the table by failing to properly document service-related conditions or by giving up after an initial denial.

Evidence-Based Research: Studies show that veterans who work with VSOs have significantly higher success rates in their VA claims compared to those who file independently (National Veterans Legal Services Program, 2020). Research also demonstrates that early medical intervention and documentation significantly improve claim outcomes and overall veteran health (VA Office of Research and Development, 2019).

Important Note: This chapter provides general guidance on VA compensation. Always consult with accredited VSO representatives or VA-accredited attorneys for advice specific to your situation. VA regulations and procedures can change, so verify current requirements with official VA sources.

Notes

Military to Civilian Relationships

Chapter

10

Transitioning from military to civilian life involves more than just changing careers - it requires navigating fundamental shifts in how relationships work. The bonds forged in military service, characterized by shared sacrifice, clear hierarchies, and common purpose, operate differently from civilian relationships. This chapter explores how to maintain meaningful military connections while building new civilian relationships and how to bridge the gap between these two worlds successfully.

Understanding Military vs. Civilian Relationship Dynamics

Military relationships are forged in unique circumstances that create deep bonds rarely found in civilian life. Understanding these differences is crucial for successful transition and relationship building in civilian contexts.

Military Relationship Characteristics:

- **Shared sacrifice** creates instant understanding and respect
- **A clear hierarchy** provides structure for interactions and decision-making
- **Common mission** aligns everyone toward shared objectives
- **Mutual dependence** where each person's performance affects everyone's safety and success
- **Unspoken understanding** about service experiences and military culture
- **Brotherhood/sisterhood** that transcends individual differences
- **Trust built under pressure** creates lasting loyalty

Civilian Relationship Characteristics:

- **Individual achievement** often takes precedence over group success
- **Informal hierarchy** with fluid leadership based on expertise or situation
- **Personal goals** that may or may not align with others' objectives
- **Conditional support** that depends on personal benefit or compatibility
- **Explicit communication** is required since shared experience can't be assumed
- **Friendship by choice** rather than necessity or assignment
- **Trust built gradually** through consistent positive interactions over time

The Challenge of Relationship Translation

One of the most difficult aspects of veteran transition is learning to build meaningful relationships in a civilian world that operates by different rules. Military veterans often struggle with civilian relationships that seem shallow, transactional, or unreliable compared to military bonds.

Common Veteran Relationship Challenges:

- **Civilian relationships feel superficial** compared to military bonds
- **Difficulty trusting civilians** who haven't proven themselves under pressure
- **Impatience with civilian problems** that seem trivial compared to military experiences
- **Communication style mismatches** between direct military communication and diplomatic civilian interaction
- **Different loyalty expectations** leading to disappointment when civilians don't show military-level commitment
- **Social fatigue** from constantly explaining military experiences or translating military references

Maintaining Military Relationships in Civilian Life

Your military relationships are valuable assets that should be maintained and nurtured throughout your civilian life. These connections provide understanding, support, and networking opportunities that civilian relationships often cannot match.

Strategies for Maintaining Military Connections:

1. Veteran Organizations and Groups:

- **Formal veteran organizations** like VFW, American Legion, or branch-specific groups
- **Informal veteran meetups** in your local area or professional field
- **Online veteran communities** that maintain connections across distances
- **Veteran networking events** for professional and social connections

2. Regular Communication:

- **Scheduled check-ins** with close military friends to maintain connection
- **Social media engagement** to stay updated on military friends' civilian transitions

- **Annual reunions** or gatherings to maintain unit or deployment connections
- **Mentorship relationships** with both junior and senior veterans

3. Shared Activities:

- **Veteran service projects** that combine social connection with meaningful service
- **Sports or recreational activities** that appeal to military camaraderie
- **Professional development** through veteran-focused career events
- **Family activities** that include military families in civilian contexts

Building Meaningful Civilian Relationships

While maintaining military connections is important, building civilian relationships is essential for successful transition and integration into civilian communities. This requires adapting your relationship-building approach to civilian contexts.

Understanding Civilian Relationship Building:

Civilian relationships typically develop through:

- **Shared interests** rather than shared danger or mission
- **Gradual trust building** over multiple positive interactions
- **Mutual benefit** where both parties gain from the relationship
- **Social compatibility,** including similar values, humor, or lifestyle
- **Geographic proximity** and regular contact opportunities
- **Professional networking** through work, industry, or career connections

Strategies for Building Civilian Relationships:

1. Interest-Based Connection:

- **Join civilian groups** based on hobbies, sports, or interests you enjoy
- **Volunteer for causes** you care about to meet like-minded civilians
- **Take classes or workshops** to learn new skills while meeting people
- **Attend community events** to become familiar with local civilian culture

2. Professional Networking:

- **Board of Directors** positions with local non-profits and charter schools
- **City- or County-wide Leadership Development Programs** that connect you with political and business leaders in your area (i.e., www.LeadershipTampaBay.com, etc.)
- **Industry associations** related to your civilian career field
- **Professional development events** to build career-focused relationships
- **Workplace social activities** to connect with colleagues outside of work
- **Business networking groups** to develop professional relationships

3. Family and Community Integration:

- **School activities,** if you have children, connect with other parents
- **Neighborhood involvement** through HOAs, community groups, or local events
- **Religious or spiritual communities** that align with your beliefs
- **Recreational leagues** for sports or competitive activities

Bridging Military and Civilian Relationships

Successfully integrating into civilian life doesn't mean abandoning your military identity or relationships. The goal is to create bridges between your military past and civilian future that honor both experiences.

Strategies for Integration:

1. Translation and Education:

- **Explain military experiences** in ways civilians can understand and appreciate
- **Share stories** that highlight universal values like teamwork, courage, and service
- **Educate civilians** about military culture without being condescending
- **Find common ground** in shared values, goals, or experiences

2. Mixed Social Groups:

- **Organize events** that include both military and civilian friends
- **Create opportunities** for your different social circles to interact
- **Facilitate introductions** between military and civilian friends with common interests
- **Host gatherings** that showcase both your military pride and civilian interests

3. Value-Based Connection:

- **Identify shared values** between military and civilian friends
- **Focus on character** rather than background when building relationships
- **Appreciate different perspectives** that civilians bring to your life
- **Maintain authenticity** while adapting your communication style

Navigating Relationship Challenges During Transition

The transition period can strain existing relationships while making it difficult to form new ones. Understanding common challenges helps you navigate these difficulties more effectively.

Common Relationship Challenges:

Family Relationship Strain:

- **Reintegration difficulties** after deployments or military separations
- **Role changes** as you transition from military member to civilian family member
- **Communication gaps** between military experiences and family understanding
- **Identity shifts** affecting how you relate to family members
- **Returning to your hometown** after being gone for 10+ years, and they don't really know you as a mature adult or how to interact with your children, who may have more lived experience than their children

Strategies for Family Relationship Healing:

- **Open communication** about your transition challenges and needs
- **Family counseling** to address military-related relationship issues
- **Patience** as family members adjust to your post-military identity
- **Shared activities** that help rebuild civilian family connections
- **Attend as many extended family events as possible** to ease your new spouse/ children into their lives and avoid being offended if they forgot to send your family an invitation; it's usually an oversight

Romantic Relationship Considerations:

- **Military relationships** may face challenges if both partners are transitioning
- **Civilian partner education** about military culture and transition challenges
- **Dating as a veteran** requires explaining military background to new partners
- **Communication style adaptation** for civilian romantic relationships

Workplace Relationship Dynamics:

- **Authority adjustment** when military rank doesn't translate to civilian status...most of the staff won't care about your former military rank, they care about how you treat them
- **Team dynamics** that differ from military unit cohesion
- **Professional networking** that requires different skills than military relationship building
- **Mentor relationships** with civilians who may not understand military experience

Incorporating the Emotional Bucket Concept

Understanding how your Emotional Bucket affects relationships is crucial during transition. When your bucket is full of transition stress and unprocessed military experiences, it becomes difficult to form new relationships or maintain existing ones effectively.

Impact of a Full Emotional Bucket on Relationships:

- **Reduced patience** with civilian relationship's pace and style
- **Communication barriers** when emotions overwhelm rational interaction
- **Trust difficulties** when past military experiences create emotional walls
- **Social withdrawal** when relationship building feels emotionally exhausting

Benefits of Managing Your Emotional Bucket:

- **Enhanced empathy** for both military and civilian relationship styles
- **Improved communication** that bridges military and civilian cultures
- **Increased patience** for building civilian relationships gradually
- **Better boundaries** that protect both military and civilian relationships

Exercise

RELATIONSHIP MAPPING AND PLANNING

Objective: Assess your current relationships and create a plan for maintaining military connections while building civilian relationships.

1. Current Relationship Assessment

Military Relationships:

Name	Relationship Type	Contact Frequency	Relationship Quality	Maintenance Needed
____________________	Battle Buddy	Weekly/Monthly/ Rarely	Strong/Good/ Weak	Y/N
____________________	Former Unit Member	Weekly/Monthly/ Rarely	Strong/Good/ Weak	Y/N
____________________	Military Mentor	Weekly/Monthly/ Rarely	Strong/Good/ Weak	Y/N
____________________	Military Friend	Weekly/Monthly/ Rarely	Strong/Good/ Weak	Y/N

Civilian Relationships:

Name	Relationship Type	Contact Frequency	Relationship Quality	Development Needed
____________________	Work Colleague	Weekly/Monthly/ Rarely	Strong/Good/ Weak	Y/N
____________________	Neighbor	Weekly/Monthly/ Rarely	Strong/Good/ Weak	Y/N
____________________	Interest-Based Friend	Weekly/Monthly/ Rarely	Strong/Good/ Weak	Y/N
____________________	Professional Contact	Weekly/Monthly/ Rarely	Strong/Good/ Weak	Y/N

2. Relationship Goal Setting

Military Relationship Goals:

- Maintain connection with _____ key military relationships
- Participate in ______ veteran organizations or activities
- Attend ____ military reunions or gatherings annually
- Mentor ____ transitioning veterans

Civilian Relationship Goals:

- Build ____ new civilian friendships this year
- Join ______ civilian organizations or groups
- Develop _____ professional relationships in my field
- Integrate into ______ community activities

3. Bridge-Building Plan

Integration Opportunities:

- Host _______________ mixed military/civilian social events
- Introduce military friends to civilian interests: ____
- Share military experiences with civilian friends through: _____________________________
- Create opportunities for military and civilian friends to meet: _________________________

4. Communication Strategy

For Military Relationships:

- Maintain communication through: _____________________________
- Share civilian transition experiences: _____________________________
- Seek support for transition challenges: _____________________________

For Civilian Relationships:

- Explain military background by: _____________________________
- Find common ground through: _____________________________
- Adapt communication style by: _____________________________

Self-Reflective Questions

1. How do I feel about the differences between military and civilian relationships, and what adjustments do I need to make?

2. What military relationships are most important to maintain, and how can I ensure they remain strong during my civilian transition?

3. What barriers prevent me from building meaningful civilian relationships, and how can I address them?

4. How can I honor my military identity while building authentic civilian relationships?

5. What role does my Emotional Bucket play in my ability to form and maintain both military and civilian relationships?

Affirmations

1. I value both my military relationships and my growing civilian connections.
2. I can build meaningful relationships in civilian life while honoring my military bonds.
3. My military experience provides strength for building relationships, not barriers.

Actionable Steps

1. **Audit Your Relationships**: Complete the relationship mapping exercise to understand your current relationship landscape.
2. **Join One New Group**: Identify and join one civilian organization or group based on your interests or career goals.
3. **Schedule Regular Contact**: Set up regular communication with three important military relationships to maintain connection.
4. **Practice Bridge-Building**: Plan one event or activity that brings together your military and civilian friends.
5. **Develop Communication Skills**: Practice explaining your military experience to civilians in ways they can understand and appreciate.

Building Your Relationship Support Network

Success in civilian transition often depends on having a strong support network that includes both military and civilian relationships. By maintaining valuable military connections while building meaningful civilian relationships, you create a comprehensive support system that serves all aspects of your post-military life.

Remember that building civilian relationships takes time and patience. Don't expect to immediately replicate the deep bonds formed in military service. Instead, focus on building authentic connections based on shared values, interests, and mutual respect. Your military experience provides valuable qualities for relationship building - loyalty, integrity, and commitment - that civilians often appreciate once they understand them.

Evidence-Based Research: Studies show that veterans who maintain both military and civilian social connections report better mental health outcomes and more successful transitions compared to those who rely solely on one type of relationship (Sayer et al., 2010). Research also indicates that veterans who actively work to bridge military and civilian cultures experience reduced isolation and improved community integration (Ahern et al., 2015).

Soft Landing – Cultivating Compassion

Chapter 11

Civilian transition often presents veterans with a myriad of emotional challenges and upheavals. Understanding the importance of being a "soft landing" - a source of empathy, compassion, and unconditional support - is crucial for veterans as they navigate new relationships and community connections. This role involves more than just understanding someone's pain; it's about offering reassurance and presence, communicating to a friend, family member, or fellow veteran, "grab my hand, I've got you," or "I may not fully grasp your pain, but I'm here to support you."

For many veterans, providing this type of support comes naturally. Military service often develops individuals who are more attuned to others' struggles, intuitively knowing how to respond sensitively to emotional needs. These veterans create safe spaces where their significant other, civilian friends, family members, and fellow veterans feel understood and supported.

Conversely, some veterans with more direct, mission-focused personalities might struggle with providing soft landings. Military training emphasizes problem-solving and direct action, which can translate to a fix-it attitude that may seem dismissive or insensitive to someone in emotional distress. Common well-intentioned but misguided responses such as, "At least you still have your health," or "You can always find another job," might be logical but can minimize the feelings of others.

Developing the ability to be a soft landing for veterans with direct personalities involves learning to practice empathy and patience. It requires recognizing and validating the feelings of others without rushing to solve the problem or compare their struggles to military experiences.

Understanding the Military Background of Compassion

Military service develops unique perspectives on compassion and support. Veterans have often experienced situations where life and death depended on unit cohesion and mutual support. This background provides both advantages and challenges in civilian compassionate care.

Military-Developed Compassion Strengths:

- **Deep loyalty** and commitment to supporting others
- **Understanding of sacrifice** and its impact on individuals and families
- **Crisis response skills** for supporting others during emergencies
- **No-judgment approach** toward those facing difficult circumstances
- **Protective instincts** that drive desire to help and shield others from harm
- **Shared experience understanding** with fellow veterans facing similar transitions

Potential Military-to-Civilian Compassion Challenges:

- **Problem-solving focus** that may overlook emotional processing needs
- **Stoicism** that may minimize others' emotional expressions
- **Comparative perspective** that may undervalue civilian struggles
- **Direct communication style** that needs softening for sensitive situations
- **Action orientation** that may rush past necessary emotional processing

Two Approaches to Compassionate Support

The Intuitive Approach (Naturally Compassionate Veterans): Some veterans naturally understand how to provide emotional support. These individuals typically:

- **Listen without judgment** to understand rather than to respond
- **Validate emotions** without trying to immediately fix problems
- **Offer presence** rather than solutions as the primary form of support
- **Ask open-ended questions** that allow others to express their feelings
- **Provide comfort** through their understanding and empathy

The Learned Approach (Direct-Style Veterans): Veterans with more direct personalities can develop compassionate skills by:

- **Practicing active listening** techniques that focus on understanding
- **Learning to validate emotions** before offering practical solutions
- **Asking about feelings** rather than immediately jumping to action plans
- **Slowing down responses** to ensure the person feels heard before being advised
- **Developing emotional vocabulary** to better connect with others' experiences

Methods for Developing Soft Landing Skills

For Naturally Compassionate Veterans:

1. **Deepen Empathy Skills:**
    - Practice mindfulness to stay present with others' emotions

- Develop cultural competence to understand diverse civilian experiences
- Learn about different types of trauma and their effects
- Study active listening techniques to enhance natural abilities

2. **Structured Follow-Up:**
    - Establish systems for checking in on those you support
    - Create consistent availability for friends and family in need
    - Develop relationships with professional resources for referrals when needed
    - Build networks of support that extend beyond your individual capacity

3. **Resource Awareness:**
    - Become knowledgeable about local mental health resources
    - Learn about veteran-specific and civilian support networks
    - Understand when to provide support versus when to refer to professionals
    - Develop connections with counselors, therapists, and support groups

For Direct-Style Veterans:

1. **Active Listening Skill Development:**
    - Practice focusing completely on the speaker without planning your response
    - Learn to ask clarifying questions that demonstrate understanding
    - Develop patience for emotional processing that may seem inefficient
    - Practice reflecting back what you hear to confirm understanding

2. **Emotional Intelligence Training:**
    - Study emotional recognition in yourself and others
    - Learn about different emotional responses to stress and trauma
    - Practice identifying emotions in facial expressions and body language
    - Develop vocabulary for discussing emotions beyond basic categories

3. **Empathy Practice:**
    - Make conscious efforts to understand perspectives different from your own
    - Practice validating emotions even when you don't fully understand them
    - Learn to sit with discomfort when you can't immediately fix someone's problem
    - Develop phrases that show empathy without minimizing experiences

Where People Seek Support: Suitable vs. Unsuitable Landing Places

When individuals are struggling emotionally, they naturally seek support from various sources. Understanding the difference between healthy and unhealthy support-seeking can help you guide people toward positive resources.

Suitable Support Sources (Healthy Soft Landings):

- **Supportive family members** who provide understanding and acceptance (i.e., parent, sibling, cousin, etc.)
- **Trusted friends** who offer emotional support without judgment
- **Mental health professionals** trained to provide clinical support
- **Spiritual leaders or communities** that align with personal beliefs
- **Support groups** for people facing similar challenges
- **Mentors or role models** who have successfully navigated similar difficulties
- **Fellow veterans** who understand military-related challenges

Unsuitable Support Sources (Hard Landings):

- **Bars and alcohol** as primary coping mechanisms (i.e., happy hour after work, etc.)
- **Gambling establishments** that provide temporary escape but create additional problems
- **Shopping or spending** for emotional relief that creates financial stress
- **Clubs or party environments** that avoid rather than address underlying issues
- **Tobacco or Substance use** including drugs or excessive prescription medication

- **Risky behaviors** that provide adrenaline but increase danger (includes sex addiction)
- **Toxic relationships or cheating on your spouse** including reconnecting with former romantic partners when currently committed to someone else
- **Gaming addiction** includes spending three or more hours daily on a video game console
- **Sugar and junk food addiction** include chips, chocolate, cake, cookies, ice cream, etc.

Helping Others Find Healthy Support

As a veteran developing soft landing skills, you can help guide others toward healthy support sources:

Recognition of Warning Signs:

- **Increased substance use** as a coping mechanism
- **Social isolation** and withdrawal from healthy relationships
- **Risky behavior patterns** that seem designed to avoid emotional processing
- **Financial problems** related to escape behaviors
- **Relationship conflicts** arising from poor coping choices

Gentle Guidance Strategies:

- **Express concern** without judgment about harmful coping patterns
- **Offer alternatives** by suggesting healthier support options
- **Provide information** about professional resources when appropriate
- **Model healthy coping** through your own behavior and choices
- **Maintain connection** even when someone makes poor support choices

The Veteran's Role as Community Soft Landing

Veterans often find themselves in natural leadership and support roles within their civilian communities. Your military experience with supporting others under pressure translates well to providing civilian emotional support when adapted appropriately.

Community Support Roles for Veterans:

- **Mentoring youth** can be very rewarding, and it will remind you of mentoring your former military subordinates, providing a tremendous sense of purpose
- **Crisis responders** in emergency situations
- **Support network builders** who connect people with resources
- **Advocacy leaders** for community improvement
- **Bridge builders** between different community groups

Mutual Growth Opportunities Between Veteran Types

Both naturally compassionate (heart-strong) and direct-style (head-strong) veterans can benefit from learning from each other:

Joint Learning Activities:

- **Compassion workshops** that combine natural empathy with strategic thinking
- **Peer feedback sessions** where veterans share approaches and learn from each other
- **Mentoring partnerships** between heart-strong and head-strong veterans
- **Role-playing exercises** that allow practice of different support approaches
- **Community service projects** that provide opportunities to practice compassionate leadership

Cross-Training Benefits:

- **Heart-strong veterans** can learn strategic problem-solving approaches from head-strong veterans
- **Headstrong veterans** can develop emotional sensitivity from naturally heart-strong veterans
- **Both approaches** can be valuable in different situations and with different people
- **Balanced support** emerges when veterans can flex between approaches as needed

Incorporating the Emotional Bucket Concept

Understanding how your own Emotional Bucket affects your ability to provide support is crucial. When your bucket is full, you may struggle to be present for others or may unconsciously project your own emotional state onto those seeking support.

Impact of a Full Emotional Bucket on Compassionate Support:

- **Reduced emotional availability** for others' needs
- **Projection of personal struggles** onto others' situations
- **Impatience with others' emotional processing** when you're overwhelmed
- **Difficulty maintaining boundaries** between your emotions and others'

Benefits of Managing Your Emotional Bucket:

- **Enhanced emotional availability** to truly support others
- **Clearer perception** of others' needs versus your own emotional reactions
- **Increased patience** for others' emotional processing and healing journeys
- **Better boundaries** that protect both you and those you're supporting

Exercise

PRACTICING COMPASSION IN REAL LIFE

Objective: This exercise aims to develop veterans' ability to offer compassionate support, enhancing empathy and understanding in real-life scenarios while honoring military experience.

Instructions:

1. **Role-Playing Sessions:** Partner with a fellow veteran or trusted civilian friend. Each pair will alternate between being a 'supporter' and a 'person seeking support.' The person seeking support will describe a challenging situation they are comfortable sharing, while the supporter practices providing a soft landing.

2. **Feedback Exchange:** After each role-playing session, provide constructive feedback on how effectively the supporter demonstrated empathy and understanding without immediately jumping to problem-solving.

3. **Scenario Practice:** Practice different scenarios including:

   - Civilian job loss or career challenges
   - Family relationship difficulties
   - Financial stress or uncertainty
   - Health concerns or medical issues
   - Social isolation or loneliness
   - Fellow veteran transition struggles

4. **Communication Skill Development:** Practice specific phrases and approaches:

   - "That sounds really difficult. Can you tell me more about how you're feeling?"
   - "I can see this is really affecting you. What would be most helpful right now?"
   - "I may not have experienced exactly what you're going through, but I'm here to listen."
   - "Would you like me to just listen, or would you like suggestions?"

5. **Group Discussion:** Hold a group discussion to share insights and personal reflections on the experience, focusing on:
    - What felt natural versus challenging about providing support
    - How military experience helped or hindered compassionate response
    - Strategies for adapting military support skills to civilian contexts

Reflection Questions for Individual Consideration:

- How does my military background influence my approach to supporting others?
- What situations trigger my desire to immediately fix problems rather than provide emotional support?
- How can I better balance my natural problem-solving skills with emotional presence?
- What personal boundaries do I need to maintain while supporting others?

Self-Reflective Questions

1. **Understanding Personal Barriers:** What personal barriers prevent me from being a compassionate supporter, and how might my military experience contribute to these barriers?

2. **Assessing Natural Tendencies:** Do I tend to listen and empathize first, or do I immediately jump to offering solutions? How can I adapt this tendency for different situations?

3. **Impact Assessment:** When I provide support to others, do I feel I make a positive impact? How can I tell, and what feedback have I received?

4. **Learning from Experience:** What have I learned from instances where I might not have provided adequate support, and how can I apply these lessons?

5. **Improvement Opportunities:** What specific aspects of compassion and support can I improve based on feedback from others and self-reflection?

Affirmations

1. **Empathy Affirmation:** "I am fully present and attentive to the needs of others, offering my support without judgment while honoring my military values."

2. **Growth Affirmation:** "Every interaction is an opportunity to learn and grow in my ability to be compassionate while maintaining my authentic self."

3. **Connection Affirmation:** "I create meaningful connections through my empathy and understanding, building bridges between my military experience and civilian community."

Actionable Steps

1. **Daily Intention Setting:** Begin each day by setting an intention to be mindful of opportunities to offer support to others in your civilian community.

2. **Active Listening Practice:** Engage in at least one conversation daily where you focus entirely on understanding the other person's perspective without interrupting or immediately offering advice.

3. **Empathy Journal:** Keep a journal where you record instances of providing support, noting what went well, what could be improved, and how your military background influenced your approach.

4. **Skill Development:** Attend workshops or training sessions on emotional intelligence, active listening, or civilian cultural competence to enhance your support skills.

5. **Community Engagement:** Volunteer in roles that require offering emotional support, such as veteran mentoring programs, community helplines, or peer support groups, to practice and deepen your compassion skills while serving others.

By developing skills as a soft landing for others, veterans can continue their service mission in civilian life while building meaningful relationships and contributing positively to their communities. The goal is to adapt the protective and supportive instincts developed in military service to civilian contexts where emotional support and empathy are often more valued than immediate problem-solving.

Evidence-Based Research: Research indicates that veterans who engage in supportive roles within their civilian communities report higher levels of purpose and life satisfaction during transition (Whiteman & Barry, 2011). Studies also show that developing empathy and emotional intelligence skills significantly improves both personal relationships and professional success in civilian environments (Mayer & Salovey, 1997).

Notes

Building Your Civilian Support Network and Overcoming Isolation

Chapter

12

The transition from military to civilian life presents unique challenges that require a strong support network to navigate successfully. In military service, your support network was built in through unit structure, shared missions, and the brotherhood or sisterhood of service. In civilian life, you must intentionally build a new support network that understands your veteran identity while helping you thrive in civilian contexts, while simultaneously overcoming the isolation that many veterans experience during this critical transition period.

Understanding Military vs. Civilian Support Networks

Military support networks are fundamentally different from civilian ones in structure, purpose, and function. Understanding these differences helps you build more effective civilian support systems while recognizing why isolation can feel so profound during transition.

Military Support Network Characteristics:

- **Automatic membership** through unit assignment and military service
- **Shared experience** of military culture, training, and service
- **Mutual dependence** where individual success affects unit success
- **Clear hierarchy** providing structure for support and guidance
- **Institutional backing** with formal support systems and resources
- **Geographic concentration** with most support network members in proximity
- **Mission-focused relationships** built around accomplishing military objectives
- **24/7 availability** with battle buddies always accessible
- **Unconditional commitment** forged through shared sacrifice and trust

Civilian Support Network Characteristics:

- **Intentional relationship building** requires active effort to create connections
- **Diverse backgrounds** with varying life experiences and perspectives
- **Individual focus** where personal success is primarily individual responsibility
- **Informal structure** with fluid leadership and support roles
- **Personal initiative** required to access support and resources
- **Geographic dispersion** with support network members often scattered

- **Relationship-focused connections** built around personal compatibility and mutual benefit
- **Scheduled availability** with planned interactions and boundaries
- **Conditional support** is often based on reciprocity and mutual benefit

The Reality of Veteran Isolation

Many veterans experience profound isolation during civilian transition, often feeling caught between two worlds - no longer fully part of the military community but not yet integrated into civilian society. This isolation can manifest in several ways:

Common Isolation Experiences:

- **Cultural disconnect** from civilians who don't understand military experience
- **Loss of identity** when military rank and role no longer define you
- **Communication barriers** when military directness doesn't translate well
- **Trust issues** when civilian relationships feel shallow compared to military bonds
- **Social anxiety** in civilian settings that feel unfamiliar
- **Geographic displacement** when moving away from military communities
- **Employment challenges** that affect social connections through work
- **Family strain** as relationships adjust to your post-military identity

Why Veteran Isolation Occurs:

- **Shared experience gap** - Civilians haven't experienced military service
- **Communication style differences** - Military directness vs civilian diplomacy
- **Different life priorities** - Service before self vs individual achievement
- **Trust-building differences** - Earned under pressure vs built gradually
- **Social ritual differences** - Military traditions vs civilian customs
- **Purpose and mission loss** - From a clear military mission to an unclear civilian purpose

The Importance of Civilian Support Networks for Veterans

Building a strong civilian support network is crucial for several reasons that go beyond simple social connections. These networks provide the foundation for successful long-term civilian integration.

Benefits of Strong Civilian Support Networks:

- **Career advancement** through professional networking and mentorship opportunities
- **Community integration** helps you feel connected to your civilian environment
- **Emotional support** during challenging transition periods and life changes
- **Practical assistance** with civilian life navigation and problem-solving
- **Social fulfillment,** providing meaningful relationships and activities
- **Resource access** connecting you to opportunities and services you might not find alone
- **Identity development** supporting your growth as a civilian while honoring your military service
- **Isolation prevention,** providing regular social contact and a sense of belonging
- **Mental health support** reducing depression and anxiety, which are common during transition
- **Family integration** helps military families connect with civilian communities

Types of Support You Need in Civilian Life

Effective civilian support networks include different types of support that serve various aspects of your post-military life. Understanding these types helps you intentionally build comprehensive support systems.

Professional Support:

- **Career mentors** who understand your industry and can guide professional development
- **Industry contacts** who can provide job opportunities and professional advice
- **Skill development partners** who help you learn civilian workplace dynamics

- **Business network connections** for entrepreneurship or business development opportunities
- **Professional references** who can vouch for your abilities in civilian contexts
- **LinkedIn connections** for online professional networking
- **Trade associations** relevant to your civilian career field

Emotional and Social Support:

- **Trusted confidants** who understand your transition challenges
- **Social companions** for recreational activities and friendship
- **Family support** including spouse, children, and extended family
- **Spiritual support** through religious or spiritual communities that align with your beliefs
- **Crisis support** for times when you need immediate emotional assistance
- **Mental health professionals** familiar with veteran issues
- **Peer counselors** who have navigated similar transitions

Practical Support:

- **Local expertise** from people who know your geographic area well
- **Service providers,** including healthcare, financial, and legal professionals
- **Community connections** who can help you navigate local systems and resources
- **Emergency support** for practical needs during difficult times
- **Information sources** that keep you informed about opportunities and resources
- **Transportation networks** for getting around your new community
- **Childcare support** for veteran families with children

Veteran-Specific Support:

- **Fellow veterans** who understand military culture and transition challenges
- **Veteran service organizations** that provide resources and advocacy

- **Military family networks** that understand the unique aspects of military life
- **Veteran mentors** who have successfully navigated civilian transition
- **Military unit connections** maintained from your service experience
- **VA-connected networks** through healthcare and benefits services
- **Combat veteran groups** for those with deployment experience

Strategies for Building Your Civilian Support Network

Building an effective civilian support network requires intentional effort and strategic thinking. Unlike military networks that developed naturally through service, civilian networks must be cultivated through deliberate action.

Professional Network Building:

1. **Industry Research and Engagement:**
    - Identify key professional organizations in your chosen civilian field
    - Attend industry conferences, meetings, and networking events
    - Join online professional groups and participate in discussions
    - Seek informational interviews with established professionals
    - Volunteer for industry-related causes and projects
2. **Workplace Relationship Development:**
    - Invest time in getting to know colleagues beyond work tasks
    - Participate in workplace social activities and team-building events
    - Seek mentorship relationships with experienced civilian professionals
    - Offer assistance and support to colleagues to build reciprocal relationships
    - Join or create workplace affinity groups or committees
3. **Skill Development Networks:**
    - Take classes or workshops to meet others with similar learning goals
    - Join study groups or professional development cohorts

- Participate in certification programs that include networking components
- Engage with online learning communities and forums
- Find accountability partners for professional growth goals

Community and Social Network Building:

1. **Interest-Based Groups:**
    - Join clubs or organizations based on your hobbies and interests
    - Participate in sports leagues or recreational activities
    - Attend community classes or workshops on topics you enjoy
    - Volunteer for causes you care about to meet like-minded people
    - Join fitness centers or outdoor activity groups
2. **Geographic Community Integration:**
    - Attend neighborhood meetings and community events
    - Join homeowner associations or tenant organizations
    - Participate in local government or civic organizations
    - Support local businesses and become a regular customer
    - Attend cultural events and festivals in your area
3. **Family-Centered Connections:**
    - Connect with other parents through children's school activities
    - Join parent-teacher organizations or school volunteer groups
    - Participate in youth sports or activity organizations as a parent
    - Attend family-friendly community events and activities
    - Build relationships with neighbors who have families

Veteran-Specific Network Maintenance and Expansion:

1. **Formal Veteran Organizations:**
   - Join national organizations like VFW, American Legion, or DAV
   - Participate in branch-specific veteran groups (Army, Navy, Air Force, Marines, Coast Guard)
   - Engage with local veteran service organizations in your area
   - Attend veteran job fairs and networking events
   - Join veteran business organizations if you're entrepreneurially inclined
2. **Informal Veteran Connections:**
   - Maintain relationships with former unit members and military friends
   - Connect with veterans in your workplace or community
   - Join online veteran communities and social media groups
   - Attend military reunions and social gatherings
   - Participate in veteran service projects and volunteer activities
3. **Military Family Support:**
   - Maintain connections with military families from your service time
   - Connect with other military families in your civilian community
   - Participate in military family support groups and activities
   - Engage with spouse and family programs offered by veteran organizations
   - Build relationships with other veteran families with similar experiences

Overcoming Common Challenges in Network Building

Veterans often face specific challenges when building civilian support networks. Recognizing these challenges and developing strategies to address them improves your success in building meaningful connections.

Challenge: Civilian Relationships Feel Shallow

Many veterans struggle with civilian relationships that seem superficial compared to the deep bonds formed in military service.

Solutions:

- Accept that civilian relationships develop differently and more gradually
- Focus on shared values and interests rather than shared experiences
- Be patient with relationship development and invest time consistently
- Share appropriate aspects of your military experience to build understanding
- Look for civilians who demonstrate character traits you value

Challenge: Difficulty Trusting Civilian Motivations

Veterans may struggle to trust civilians who haven't proven themselves under pressure as military relationships require.

Solutions:

- Recognize that civilian trust builds differently than military trust
- Look for consistency in civilian behavior over time
- Value reliability in small matters as an indicator of larger trustworthiness
- Give people opportunities to demonstrate their character in civilian contexts
- Focus on present behavior rather than assumptions about civilian commitment levels

Challenge: Communication Style Mismatches

Military communication styles may not translate well to civilian relationship building.

Solutions:

- Practice adapting your communication style to civilian expectations
- Learn to listen more and talk less in civilian social situations
- Ask questions about others' experiences and interests
- Share your military experience in relatable ways
- Observe successful civilian communicators and adapt their approaches

Challenge: Time and Energy Constraints

Building networks requires significant time and energy investment that may feel overwhelming during transition.

Solutions:

- Start small with one or two networking activities rather than trying to do everything
- Choose activities that align with your existing interests and goals
- Set realistic expectations for relationship development pace
- Focus on quality relationships rather than quantity of connections
- Integrate networking into activities you already enjoy

Challenge: Fear of Rejection or Judgment

Veterans may fear being judged or rejected by civilians who don't understand military experience.

Solutions:

- Start with veteran-friendly environments before branching out
- Practice explaining your military background in positive, relatable terms
- Focus on what you can contribute rather than what makes you different
- Remember that many civilians respect and appreciate military service
- Build confidence through successful interactions before attempting more challenging ones

Strategies for Overcoming Isolation

Isolation during veteran transition requires active, intentional strategies to overcome. Simply building networks isn't enough - you must also address the underlying factors that contribute to feeling disconnected.

Immediate Isolation Relief Strategies:

1. **Daily Connection Rituals:**
    - Schedule at least one meaningful conversation daily
    - Send texts or emails to maintain contact with distant friends

- Participate in online communities or forums
- Call family members regularly
- Engage with neighbors or community members

2. **Structured Social Activities:**
   - Join groups with regular meeting schedules
   - Volunteer for organizations that need consistent help
   - Take classes or workshops that require attendance
   - Participate in fitness programs or sports leagues
   - Attend religious or spiritual services regularly
3. **Professional Engagement:**
   - Actively participate in workplace social activities
   - Join professional associations with local chapters
   - Attend industry meetups and networking events
   - Seek mentorship relationships
   - Offer to mentor others in your field

Long-term Isolation Prevention:

1. **Identity Integration:**
   - Develop a civilian identity that incorporates military experience
   - Find ways to use military skills in civilian contexts
   - Share your story in appropriate settings
   - Join organizations that value your military background
   - Pursue activities that align with military values
2. **Purpose Reconnection:**
   - Identify civilian activities that provide sense of mission
   - Volunteer for causes that matter to you
   - Mentor other veterans or young people

    - Engage in community improvement projects
    - Find work that feels meaningful and impactful

3. **Cultural Bridge Building:**
    - Learn about civilian culture and customs
    - Help civilians understand military experience
    - Find common ground in shared values
    - Participate in community traditions and events
    - Build relationships with other cultural bridge-builders

Maintaining and Nurturing Your Support Network

Building your network is only the beginning. Maintaining and nurturing these relationships requires ongoing effort and strategic thinking.

Regular Communication and Connection:

- Schedule regular check-ins with important network members
- Send updates about your progress and achievements to your network
- Reach out to offer support and assistance to others in your network
- Remember important events and milestones in others' lives
- Use social media strategically to maintain a connection with a broader network

Value Exchange and Reciprocity:

- Look for ways to provide value to others in your network
- Offer your military skills and experience where they can benefit others
- Make introductions between network members who could help each other
- Share resources, information, and opportunities with your network
- Be generous with recognition and appreciation for others' support

Network Evaluation and Expansion:

- Regularly assess whether your network is meeting your evolving needs
- Identify gaps in your network and actively work to fill them
- Prune relationships that are consistently negative or draining
- Expand your network as your civilian career and interests develop
- Maintain a balance between giving and receiving support

Incorporating the Emotional Bucket Concept

Your Emotional Bucket significantly affects your ability to build and maintain civilian support networks while overcoming isolation. When your bucket is full of transition stress and unprocessed military experiences, network building becomes more difficult, and isolation feels more severe.

Impact of a Full Emotional Bucket on Network Building:

- **Reduced social energy** for meeting new people and building relationships
- **Negative assumptions** about civilian motivations and reliability
- **Communication barriers** that prevent authentic connection with others
- **Impatience** with the development pace of civilian relationships
- **Social withdrawal** when networking feels emotionally exhausting
- **Increased sensitivity** to rejection or judgment
- **Heightened isolation** as emotional overload makes connection feel impossible

Benefits of Managing Your Emotional Bucket:

- **Increased openness** to new relationships and experiences
- **Enhanced social skills** for building connections with diverse people
- **Better communication** that bridges military and civilian cultures
- **Improved patience** for gradual relationship development
- **Greater emotional availability** for supporting others in your network
- **Reduced social anxiety,** making networking feel more natural
- **Enhanced empathy** for helping others feel included and supported

Exercise

SUPPORT NETWORK MAPPING AND DEVELOPMENT

Objective: Assess your current support network and create a strategic plan for building the civilian support you need for long-term success while developing specific strategies to overcome isolation.

1. Current Support Network Assessment

Military Support Network:

Name	Relationship	Support Type	Contact Frequency	Relationship Quality
____________	Unit Member	Professional/ Emotional/Social	Weekly/Monthly/ Rare	Strong/Good/ Weak
____________	Military Mentor	Professional/ Emotional	Weekly/Monthly/ Rare	Strong/Good/ Weak
____________	Military Friend	Social/Emotional	Weekly/Monthly/ Rare	Strong/Good/ Weak
____________	Veteran Colleague	Professional/ Social	Weekly/Monthly/ Rare	Strong/Good/ Weak

Civilian Support Network:

Name	Relationship	Support Type	Contact Frequency	Relationship Quality
____________	Work Colleague	Professional	Daily/Weekly/ Monthly	Strong/Good/ Weak
____________	Neighbor	Social/Practical	Weekly/Monthly/ Rare	Strong/Good/ Weak
____________	Community Member	Social/Practical	Weekly/Monthly/ Rare	Strong/Good/ Weak
____________	Professional Contact	Professional	Monthly/ Quarterly/Rare	Strong/Good/ Weak

2. Isolation Assessment

Rate your current experience (1=Never, 5=Always):

- _________I feel understood by people in my daily life
- _________I have people I can call when I need support
- _________I feel connected to my community
- _________I have meaningful conversations regularly
- _________I feel like I belong somewhere
- _________I have people who share my values
- _________I feel comfortable in social situations

Isolation Risk Factors (Check all that apply):

☐ Live far from military friends/family

☐ Work in an environment with few veterans

☐ Struggle to relate to civilian interests

☐ Feel misunderstood by civilians

☐ Avoid social situations

☐ Prefer isolation to superficial connections

☐ Feel like civilian problems are trivial

☐ Miss the camaraderie of military service

3. Support Network Gap Analysis

Professional Support Needs:

- Industry mentorship: Needed/Have/Strong
- Career networking: Needed/Have/Strong
- Skill development: Needed/Have/Strong
- Business connections: Needed/Have/Strong

Personal Support Needs:

- Emotional support: Needed/Have/Strong
- Social connections: Needed/Have/Strong
- Practical assistance: Needed/Have/Strong
- Crisis support: Needed/Have/Strong

Community Integration Needs:

- Local knowledge: Needed/Have/Strong
- Community involvement: Needed/Have/Strong
- Service opportunities: Needed/Have/Strong
- Cultural connections: Needed/Have/Strong

4. Network Building Action Plan

Professional Network Goals:

- Join ______ professional organizations this year
- Attend _____ networking events per month
- Develop ______ new professional relationships
- Seek mentorship from ______________________________ (specific type of professional)

Social Network Goals:

- Join ______ community organizations or clubs
- Participate in _____ social activities per month
- Build _____ new friendships this year
- Engage in ______ volunteer activities

Integration Activities:

- Attend _____ community events per month
- Join ______ neighborhood or local groups
- Participate in _____ family-centered activities

- Engage in ______ civic or cultural activities

5. Isolation Reduction Plan

Daily Connection Goals:

- Have ______ meaningful conversations per day
- Send ______ texts/emails to maintain distant relationships
- Participate in _____ online communities
- Engage with _____ neighbors or community members

Weekly Social Goals:

- Attend ____ group activities per week
- Volunteer ______ hours per week
- Participate in ______ structured social activities
- Make ______ new social contacts per week

Monthly Integration Goals:

- Attend _____ community events per month
- Join ______ new groups or organizations
- Host ______ social gatherings per month
- Participate in _____ cultural activities per month

6. Network Maintenance Plan

Communication Schedule:

- Weekly contact with: _____
- Monthly contact with: _____
- Quarterly contact with: _____
- Annual contact with: _____

Value Exchange Strategy:

- Skills I can offer to my network: ______
- Resources I can share: ______
- Introductions I can facilitate: ______
- Support I can provide: ______

7. Progress Tracking

Monthly Check-in Questions:

- How many new connections did I make this month?
- How often did I feel isolated this month?
- What networking activities were most successful?
- What barriers to connection did I encounter?
- How can I improve my networking approach next month?

Self-Reflective Questions

1. How does my current support network compare to what I had in the military, and what adjustments do I need to make?

2. What specific aspects of civilian relationships do I find most challenging, and how can I address these challenges?

3. When do I feel most isolated, and what patterns can I identify in these experiences?

4. What barriers prevent me from building civilian relationships, and how can I overcome them?

5. How does my Emotional Bucket affect my ability to build and maintain civilian support networks?

Affirmations

1. I am capable of building meaningful civilian relationships while maintaining my military connections.

2. My military experience provides valuable qualities for relationship building in civilian contexts.

3. I can overcome isolation by taking consistent action to connect with others.

Actionable Steps

1. **Complete Network Assessment**: Use the mapping exercise to understand your current support network strengths and gaps.

2. **Join One New Group**: Identify and join one civilian organization or group that aligns with your interests or professional goals.

3. **Schedule Regular Communication**: Set up consistent communication schedules with key network members to maintain relationships.

4. **Implement Daily Connection**: Commit to having at least one meaningful interaction with another person every day.

5. **Create Isolation Prevention Plan**: Develop specific strategies for recognizing and addressing isolation before it becomes overwhelming.

Building Your Comprehensive Support System

Building your civilian support network while overcoming isolation is an investment in your long-term success and well-being. By strategically developing relationships that support your professional goals, personal growth, and community integration, you create the foundation for a fulfilling civilian life that honors your military service while embracing new opportunities and challenges.

Remember that building civilian relationships takes time and patience. Don't expect to replicate the deep bonds formed in military service immediately. Instead, focus on creating authentic connections based on shared values, interests, and mutual respect. Your military experience provides valuable qualities for relationship building - loyalty, integrity, and commitment - that civilians often appreciate once they understand them.

The goal is not to choose between military and civilian relationships, but to create a comprehensive support network that includes both. By maintaining valuable military connections while building meaningful civilian relationships, you create a support system that serves all aspects of your post-military life.

Isolation is not inevitable during veteran transition. With intentional effort and strategic relationship building, you can create a thriving social network that provides support, understanding, and connection throughout your civilian journey.

Evidence-Based Research: Studies show that veterans with strong civilian support networks report significantly better transition outcomes, including higher employment satisfaction, improved mental health, and greater community integration (Sayer et al., 2010). Research also indicates that intentional network building leads to better career advancement and life satisfaction compared to passive relationship development (Granovetter, 1995). Additionally, veterans who actively address isolation through structured social activities and professional networking report reduced symptoms of depression and anxiety during transition (Ahern et al., 2015).

Notes

SECTION III

Professional Success and Leadership

Translating Military Skills to Civilian Success

Chapter

13

One of the most significant challenges veterans face during civilian transition is effectively communicating the value of their military experience to civilian employers and colleagues. While military service develops exceptional skills and capabilities, these are often described in military terminology that civilians may not fully understand or appreciate. This chapter provides strategies for identifying, translating, and leveraging your military skills for civilian career success.

Understanding the Translation Challenge

Military service develops a comprehensive skill set that is highly valuable in civilian contexts, but the language and context of military experience can create barriers to recognition and application. The challenge isn't that your skills aren't valuable - it's that they need to be translated into civilian terms and contexts.

Common Translation Barriers:

- **Military jargon and acronyms** that civilians don't understand
- **Rank and hierarchy systems** that don't directly translate to civilian roles
- **Mission-focused descriptions** that don't highlight transferable skills
- **Tactical language** that needs strategic business translation
- **Security-related limitations** on discussing specific experiences
- **Undervaluation of leadership experience** when you managed people at a young age
- **Assumption gaps** where civilians don't understand military training depth

Why Translation Matters:

- **Career advancement** depends on the civilian understanding of your capabilities
- **Salary negotiation** requires articulating your value in business terms
- **Team integration** needs colleagues to understand your strengths
- **Leadership opportunities** require demonstrating relevant experience
- **Networking effectiveness** improves when you can connect military and civilian experiences

Core Military Skills and Their Civilian Applications

Military service develops core competencies that are highly sought after in civilian organizations. Understanding how these translate helps you position yourself effectively.

Leadership and Management:

Military Experience:

- Led squads, teams, or units under high-pressure conditions
- Managed personnel, equipment, and mission objectives
- Made critical decisions with limited information
- Developed subordinates through training and mentorship

Civilian Translation:

- Team leadership and people management
- Project management and resource allocation
- Executive decision-making under pressure
- Employee development and coaching
- Change management and organizational leadership

Problem-Solving and Critical Thinking:

Military Experience:

- Analyzed complex tactical situations rapidly
- Developed and executed operational plans
- Adapted strategies based on changing conditions
- Troubleshot equipment and system failures

Civilian Translation:

- Strategic planning and analysis
- Process improvement and optimization
- Crisis management and business continuity
- Technical troubleshooting and system optimization
- Risk assessment and mitigation planning

Communication and Coordination:

Military Experience:

- Briefed senior leaders on mission status
- Coordinated with multiple units and agencies
- Communicated effectively under stress
- Wrote detailed reports and documentation

Civilian Translation:

- Executive communication and presentation skills
- Cross-functional collaboration
- Stakeholder management
- Crisis communication
- Technical writing and documentation

Training and Development:

Military Experience:

- Designed and delivered training programs
- Mentored junior personnel
- Evaluated performance and provided feedback
- Maintained training standards and safety protocols

Civilian Translation:

- Learning and development program design
- Coaching and mentoring
- Performance management
- Quality assurance and compliance
- Safety program management

Technology and Technical Skills:

Military Experience:

- Operated sophisticated military equipment
- Maintained complex technical systems
- Learned new technologies quickly
- Adapted to rapidly changing technical environments

Civilian Translation:

- Technical aptitude and system management
- Equipment maintenance and optimization
- Rapid technology adoption
- Process automation and efficiency improvement
- Technical training and support

The STAR Method for Military Experience Translation

The STAR method (Situation, Task, Action, Result) is particularly effective for translating military experiences into compelling civilian narratives.

STAR Framework:

- **Situation**: Describe the context and challenges
- **Task**: Explain your responsibility or objective
- **Action**: Detail the specific steps you took
- **Result**: Quantify the outcomes and impact

Military Example: "Led a 12-person squad during a 15-month deployment to Afghanistan, responsible for convoy security operations covering 500+ missions with zero casualties while maintaining a 98% mission success rate."

Civilian Translation using STAR:

- **Situation:** "Managed a high-risk operational team in a challenging international environment"
- **Task:** "Responsible for ensuring team safety and mission success across 500+ critical operations"
- **Action:** "Implemented comprehensive safety protocols, conducted regular training, and maintained strict operational standards"
- **Result:** "Achieved 98% mission success rate with zero safety incidents over 15 months"
- **Civilian Impact:** Demonstrates leadership, safety management, team development, and results achievement under pressure

Industry-Specific Skill Translation

Different civilian industries value different aspects of military experience. Understanding how to position your skills for specific sectors improves your career prospects.

Corporate/Business Sector:

Valuable Military Skills:

- Leadership and team management
- Strategic planning and execution
- Process improvement and efficiency
- Risk management and compliance
- Training and development

Translation Focus:

- "Led cross-functional teams of up to X personnel"
- "Managed budgets and resources totaling $X"
- "Implemented process improvements that increased efficiency by X%"
- "Developed and executed strategic plans with measurable outcomes"
- "Trained and developed team members, resulting in X% promotion rate"

Technology Sector:

Valuable Military Skills:

- Technical aptitude and system management
- Security and risk assessment
- Rapid learning and adaptation
- Project management
- Quality assurance

Translation Focus:

- "Managed complex technical systems with 99.X% uptime"
- "Implemented security protocols protecting sensitive information"
- "Rapidly learned and deployed new technologies in high-stakes environments"
- "Led technical projects from conception to implementation"
- "Maintained quality standards in mission-critical operations"

Healthcare/Public Service:

Valuable Military Skills:

- Crisis response and emergency management
- Attention to detail and safety protocols
- Service orientation and mission focus
- Training and compliance
- Stress management

Translation Focus:

- "Managed emergency response operations under high-stress conditions"
- "Maintained strict safety and compliance standards in critical operations"
- "Demonstrated commitment to service and mission accomplishment"
- "Trained personnel in life-saving procedures and protocols"
- "Performed effectively in high-pressure, time-sensitive environments"

Manufacturing/Operations:

Valuable Military Skills:

- Operations management and logistics
- Safety and compliance
- Quality control and continuous improvement
- Equipment maintenance and optimization
- Team coordination

Translation Focus:

- "Managed complex logistics operations involving multiple moving parts"
- "Implemented safety programs resulting in zero-incident performance"
- "Led quality improvement initiatives that exceeded standards"
- "Maintained and optimized equipment for maximum efficiency"
- "Coordinated teams to achieve production and quality targets"

Overcoming Common Translation Mistakes

Veterans often make mistakes when translating their military experience that can hurt their civilian career prospects.

Common Mistakes to Avoid:

Mistake 1: Using Military Jargon

- **Wrong**: "Served as NCOIC for a 15-man squad conducting MOUT operations"
- **Right**: "Led a 15-person team conducting urban operations training and real-world missions"

Mistake 2: Focusing on Tasks Instead of Results

- **Wrong**: "Responsible for equipment maintenance"
- **Right**: "Maintained equipment achieving 98% operational readiness, exceeding standard by 15%"

Mistake 3: Undervaluing Your Experience

- **Wrong**: "Just followed orders and did my job"
- **Right**: "Executed complex operations while leading teams and managing resources under pressure"

Mistake 4: Being Too Generic

- **Wrong**: "Good leader with military experience"
- **Right**: "Proven leader who developed and managed teams of up to 25 personnel, achieving 100% retention and 95% promotion rate"

Mistake 5: Not Quantifying Impact

- **Wrong**: "Improved processes"
- **Right**: "Streamlined logistics processes, reducing delivery time by 30% and cutting costs by $50,000 annually"

Building Your Military-To-Civilian Skill Portfolio

Creating a comprehensive portfolio of translated skills helps you present yourself effectively across various civilian opportunities.

Step 1: Military Experience Inventory

Create a comprehensive list of your military experiences, including:

- Leadership positions and span of control
- Training and certifications completed
- Equipment and systems operated
- Projects managed and outcomes achieved
- Awards and recognitions received
- Deployments and special assignments

Step 2: Skill Identification and Categorization

Organize your experiences into transferable skill categories:

- Leadership and management
- Technical and analytical skills

- Communication and interpersonal skills
- Problem-solving and decision-making
- Training and development
- Project and operations management

Step 3: Civilian Language Translation

For each skill category, develop:

- Civilian terminology and descriptions
- Quantified results and achievements
- Industry-specific applications
- STAR-method stories
- Value propositions for employers

Step 4: Portfolio Validation

Test your translations by:

- Sharing with civilian colleagues or mentors
- Getting feedback from industry professionals
- Practicing in networking conversations
- Refining based on responses and reactions
- Updating as you gain civilian experience

Leveraging Military Networks for Civilian Success

Your military network can be a powerful asset for civilian career advancement when properly leveraged.

Military Network Career Resources:

- **Fellow veterans** in your target industry who can provide insights and referrals
- **Former supervisors** who can serve as references and mentors
- **Military colleagues** who have successfully transitioned to civilian careers
- **Veteran-owned businesses** that understand and value military experience

- **Military professional associations** that provide industry-specific networking

Network Leveraging Strategies:

1. **Informational Interviews**: Connect with veterans in your target field to learn about industry requirements and opportunities
2. **Referral Requests**: Ask military contacts for introductions to civilian professionals in your area of interest
3. **Mentorship Relationships**: Seek guidance from veterans who have successfully made transitions similar to yours
4. **Collaboration Opportunities**: Partner with fellow veterans on projects that demonstrate your civilian capabilities
5. **Reference Development**: Maintain relationships with military supervisors who can speak to your capabilities in civilian terms

Incorporating the Emotional Bucket Concept

Managing your Emotional Bucket is crucial for effective skill translation and civilian career success. When your bucket is full of transition stress and frustration, it becomes difficult to present yourself positively and professionally.

Impact of a Full Emotional Bucket on Skill Translation:

- **Defensive communication** when explaining military experience
- **Frustration with civilians' lack of understanding,** leading to poor interactions
- **Impatience with the translation process,** resulting in shortcuts and mistakes
- **Negative assumptions** about civilian workplace values
- **Reduced confidence** in presenting military experience effectively

Benefits of Managing Your Emotional Bucket:

- **Clear communication** that effectively bridges military and civilian contexts
- **Patient education** of civilians about the military experience value
- **Confident presentation** of skills and achievements
- **Positive attitude** that attracts opportunities and relationships
- **Adaptive approach** that adjusts to different civilian audiences

Exercise

MILITARY SKILL TRANSLATION WORKSHOP

Objective: Develop effective translations of your military experience for civilian career success.

Part 1: Experience Inventory

Leadership Experience:

- Positions held: ______________________________
- Number of people managed: ____________
- Duration of leadership roles: ____________
- Key accomplishments: ______________________________
- Challenges overcome: ______________________________

Technical Skills:

- Equipment/Systems operated: ______________________________
- Certifications earned: ______________________________
- Technical training completed: ______________________________
- Technical achievements: ______________________________

Project Management:

- Projects led or contributed to: ______________________________
- Budget/Resource responsibility: ______________________________
- Timeline achievements: ______________________________
- Measurable outcomes: ____________

Training and Development:

- Training programs delivered: ______________________________
- Personnel trained: ______________________________
- Training outcomes: ______________________________
- Curriculum development: ______________________________

Part 2: STAR Method Practice

Choose one significant military experience and complete:

Situation: Describe the context and challenges

____

____

Task: Explain your responsibility or objective

____

____

Action: Detail the specific steps you took

____

____

Result: Quantify the outcomes and impact

____

____

Part 3: Civilian Translation

Original Military Description:

____

____

Civilian Translation:

____

____

Industry-Specific Version (choose your target industry):

____

____

Value Proposition for Employers:

____

____

Part 4: Networking Elevator Pitch

Develop a 30-second pitch that includes:

- Your military background (civilian terms)
- Key transferable skills
- Value you bring to civilian organizations
- Your career interests/goals

Your Elevator Pitch:

____

____

____

____

Part 5: Translation Validation

Practice your translations with:

- A civilian colleague or friend
- A veteran who has successfully transitioned
- A professional in your target industry

Feedback received:

____

____

Adjustments needed:

____

____

Self-Reflective Questions

1. Which of my military experiences do I struggle most to explain to civilians, and how can I better translate their value?

2. What assumptions might I be making about civilian understanding of military experience that could be hindering my career progress?

3. How effectively am I quantifying the results and impact of my military experience in civilian terms?

4. What aspects of my military background provide the most value in my target civilian career field?

5. How does the state of my Emotional Bucket affect my ability to present my military experience positively and professionally?

Affirmations

1. My military experience has provided me with valuable skills that translate directly to civilian career success.

2. I can effectively communicate my military background in ways that resonate with civilian employers and colleagues.

3. I am confident in the unique value I bring to civilian organizations through my military experience.

Actionable Steps

1. **Complete Skills Inventory**: Document all your military experiences, achievements, and skills using the worksheet provided.

2. **Develop STAR Stories**: Create 5-7 compelling STAR-method stories that translate key military experiences into civilian value propositions.

3. **Practice Civilian Language**: Rewrite your resume and LinkedIn profile using civilian terminology and quantified achievements.

4. **Seek Translation Feedback**: Share your translated experience descriptions with civilian colleagues or mentors for feedback and refinement.

5. **Build Industry-Specific Versions**: Adapt your skill translations for your specific target industry and practice presenting them in networking situations.

Leveraging Your Military Foundation for Civilian Excellence

Successfully translating your military skills to civilian success requires more than just changing terminology - it requires understanding how your military experience creates value in civilian contexts. Your military service has developed capabilities that many civilians spend years trying to acquire: leadership under pressure, the ability to learn quickly, operational excellence, and mission-focused execution.

The key is presenting these skills in ways that resonate with civilian employers and colleagues while maintaining pride in your military service. You're not minimizing your military experience when you translate it - you're maximizing its impact in your new environment.

Remember that skill translation is an ongoing process. As you gain civilian experience, you'll develop new ways to connect your military background with civilian success. The goal is to become bilingual - fluent in both military and civilian professional languages - so you can serve as a bridge between these two worlds while advancing your own career.

Your military experience is not a barrier to civilian success - it's a competitive advantage that needs proper translation and presentation. With the right approach, your military background becomes a distinctive strength that sets you apart in the civilian marketplace.

Evidence-Based Research: Studies show that veterans who effectively translate their military skills into civilian terminology have significantly higher employment rates and career advancement opportunities (Corporate Gray, 2019). Research also indicates that employers value military experience when it's presented in relevant business contexts, with 89% of HR professionals rating veterans as valuable employees when their skills are properly communicated (RecruitMilitary, 2020). Additionally, veterans who develop strong skill translation capabilities report higher job satisfaction and faster career progression in civilian roles (Syracuse University Institute for Veterans and Military Families, 2018).

Notes

Notes

The F7 Neuro Mindset Method - Balanced Living

Chapter

14

The transition from military to civilian life requires a comprehensive approach to personal development that addresses all aspects of well-being. The F7 Neuro Mindset Method™, created by Dr. Roderick Cunningham, is a holistic framework explicitly designed to help veterans create balanced, fulfilling civilian lives by focusing on seven fundamental areas: Faith, Family, Finances, Fitness, Friendships, Fun, and Frequency. This method recognizes that military service provides structure and purpose across multiple life domains, and civilian success requires intentionally rebuilding that comprehensive foundation.

Understanding the F7 Neuro Mindset Method™ for Veterans

Military life provided a structured approach to personal development through rank advancement, training requirements, physical fitness standards, and clear missions. Civilian life often lacks this integrated structure, leaving veterans to create their own comprehensive development plan. Dr. Rod's F7 Neuro Mindset Method™ fills this gap by providing a systematic approach to balanced living that honors military values while embracing civilian opportunities.

Why Veterans Need the F7 Approach:

- **Comprehensive structure** replaces the integrated development system of military service
- **Mission-focused approach** provides clear objectives across life domains
- **Measurable progress** allows for the goal-oriented achievement that veterans value
- **Balance integration** prevents over-focus on career at the expense of personal well-being and family
- **Values alignment** connects civilian life with military principles of service and excellence
- **Holistic wellness** addresses the interconnected nature of physical, mental, and spiritual health

The Seven Pillars for Veteran Success

Each pillar of the F7 Method requires specific veteran adaptations to address the unique challenges and opportunities of military-to-civilian transition.

FAITH
FAMILY
FINANCES
FITNESS
FRIENDSHIP
FUN
FREQUENCY

1. Faith - Spiritual Foundation for Transition

For veterans, faith often involves processing difficult service experiences while finding new purpose and meaning in civilian life. This pillar provides stability during times of transition and connects you to something greater than yourself.

Faith in Veteran Context:

- **Purpose beyond service**: Finding meaning and mission in civilian life
- **Moral injury healing**: Processing ethical challenges from military service
- **Spiritual community**: Connecting with like-minded individuals who share your values
- **Gratitude practice**: Appreciating both military experience and civilian opportunities
- **Service continuation**: Channeling spiritual values into civilian service to others

Faith Development Strategies:

- Engage in daily spiritual practices (prayer, meditation, reflection)
- Join spiritual communities that welcome and understand veterans
- Practice forgiveness for difficult military experiences
- Find ways to serve others that honor spiritual values

Veteran Faith Considerations:

- Find faith communities that understand military experience
- Integrate military service into spiritual identity rather than separating them
- Use spiritual practices to manage transition stress and uncertainty

FAITH
YOUR ONE-ON-ONE SPIRITUAL RELATIONSHIP WITH YOUR GOD
PRAY
STUDY GOD'S WORD
MEDITATE ON HIS WORD
ASSEMBLE

2. Family - Rebuilding and Strengthening Relationships

Military service often strains family relationships through deployments, stress, and the demands of service. Civilian transition provides an opportunity to rebuild and strengthen these crucial connections while adapting to new family dynamics and prioritizing quality time with your loved ones.

Family in Veteran Context:

- **Relationship repair**: Healing relationships strained by military service
- **Role redefinition**: Adapting from military member to civilian family member
- **Communication improvement**: Learning civilian family communication after military directness
- **Presence practice**: Being fully present with family after years of mission focus
- **Legacy building**: Creating positive family traditions in civilian life
- **Quality time investment**: Dedicating focused attention to spouse and all children
- **Extended family connection**: Maintaining relationships with the broader family network

Family Development Strategies:

- **Spouse relationship priority**: Schedule regular date nights, meaningful conversations, and shared activities with your spouse
- **Individual child attention**: Spend one-on-one time with each child, regardless of custody arrangements or living situations
- **Active participation**: Attend children's sporting events, recitals, and school activities
- **Skill development**: Teach your children how to excel in sports, academics, and life skills
- **Family traditions**: Establish regular family game nights, movie nights, and special activities
- **Vacation planning**: Take family vacations and weekend trips to create lasting memories
- **Holiday engagement**: Actively participate in holiday gatherings with extended family
- **Outdoor activities**: Plan hiking, camping, fishing, or other outdoor adventures together
- **Extended family inclusion**: Maintain relationships with grandparents, siblings, aunts, uncles, and cousins

Veteran Family Considerations:

- Understand that family members may need time to trust your civilian commitment
- Address any family trauma related to military service or deployment
- Learn to share decision-making after years of independent military choices
- Balance family time with other F7 areas without guilt
- Make extra effort to connect with children who don't live with you full-time
- Include extended family in your civilian life to maintain important connections
- Create new family memories that help heal any wounds from military absence

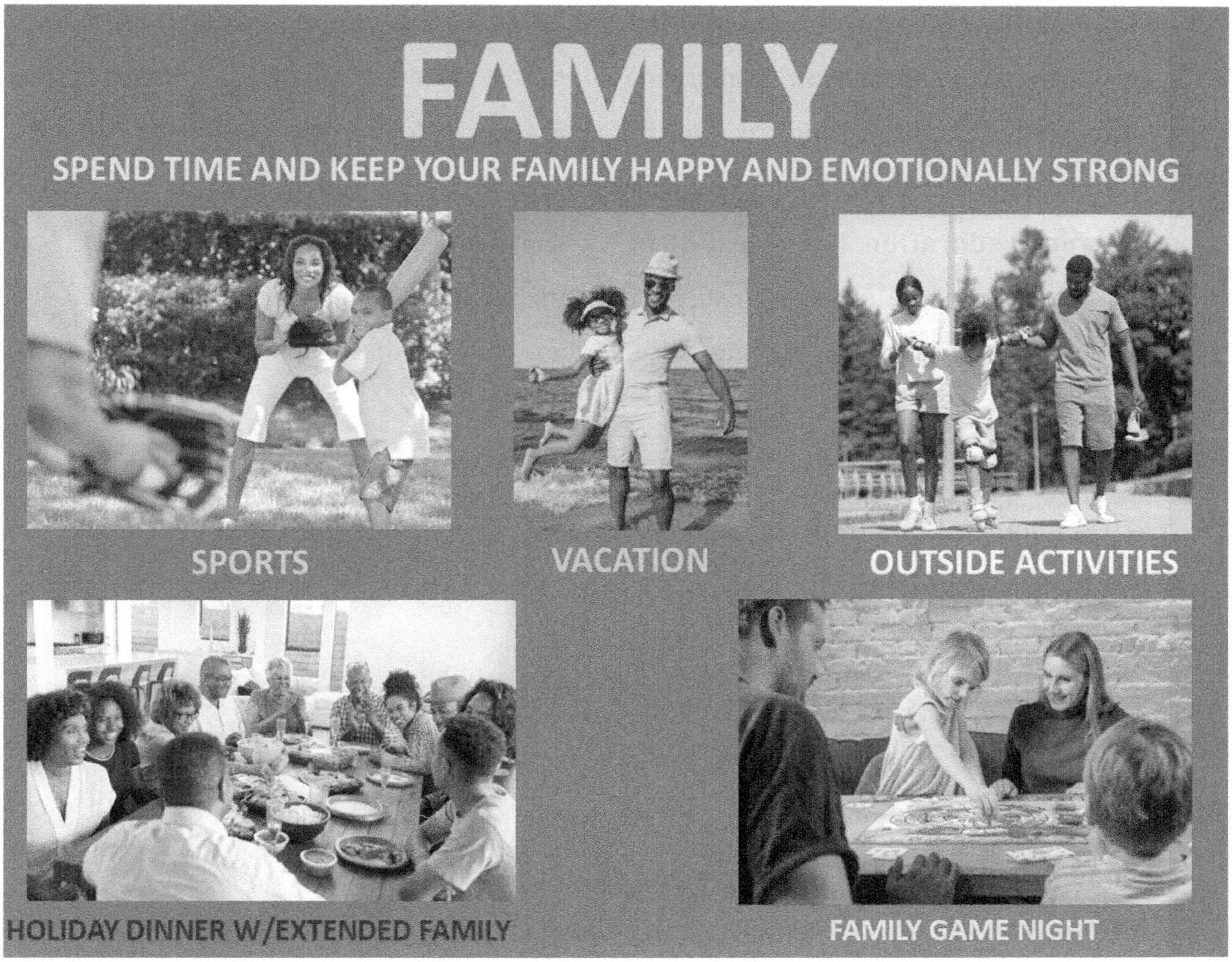

3. Finances - Building Financial Security

Military service provided a financial structure through guaranteed pay, benefits, and clear advancement, including basic necessities like housing, food, and healthcare. Civilian financial management requires new skills and approaches while maximizing veteran-specific benefits and opportunities, with the added responsibility of providing all basic needs for yourself and your family. You are 100% accountable for the financial security of your family. No one is going to come and rescue you and your family.

Finances in Veteran Context:

- **Basic needs provision**: Taking responsibility for food, clothing, and shelter for your family
- **VA benefits optimization**: Maximizing disability, education, and healthcare benefits
- **Civilian career transition**: Translating military experience into optimal compensation
- **Emergency fund building**: Creating financial security for career uncertainty
- **Investment strategy**: Building wealth beyond military retirement benefits
- **Financial education**: Learning civilian financial management and investment
- **Transportation reliability**: Ensuring dependable vehicles for family needs
- **Career responsibility**: Making wise career choices that support family financial stability

Financial Development Strategies (20-60-20 Method):

- **20% Financial Planning**: Savings, investments, debt payment, emergency fund
- **60% Non-Negotiable Expenses**: Housing, food, transportation, insurance, utilities, clothing
- **20% Discretionary Spending**: Entertainment, hobbies, non-essential purchases

Essential Family Financial Responsibilities:

- **Housing security**: Ensure stable, safe housing for your family
- **Food provision**: Budget for nutritious meals and grocery needs
- **Clothing needs**: Provide appropriate clothing for all family members
- **Transportation**: Maintain reliable vehicles for work, school, and family activities
- **Healthcare coverage**: Secure adequate medical coverage for family health needs
- **Education funding**: Plan for children's educational expenses and opportunities

Veteran Financial Considerations:

- Understand how VA benefits integrate with civilian financial planning
- Leverage veteran-specific financial benefits (VA home loans, education benefits)
- Plan for potential career changes during civilian transition, but avoid impulsive entrepreneurship ventures
- Build civilian credit and financial relationships
- Take full responsibility for family financial security, which the military previously provided
- Make career decisions based on family financial needs, not just personal preferences
- Seek experienced mentorship before making major entrepreneurial decisions that could jeopardize family security

4. Fitness - Adapting Military Fitness to Civilian Wellness

Military fitness is focused on mission readiness and unit standards. Civilian fitness should focus on long-term health, stress management, and personal well-being while maintaining the discipline and excellence of military training. This includes comprehensive health management beyond just exercise.

Fitness in Veteran Context:

- **Mission adaptation**: Shifting from military readiness to civilian wellness
- **Injury management**: Addressing service-related injuries while maintaining activity
- **Stress relief**: Using exercise to manage transition and civilian life stress
- **Identity maintenance**: Preserving military fitness identity in civilian contexts
- **Community building**: Finding civilian fitness communities that respect military background
- **Comprehensive health**: Taking full responsibility for your physical and mental health
- **Preventive care**: Maintaining regular medical checkups and health monitoring

Fitness Development Strategies:

- **Annual health physicals**: Schedule and attend yearly comprehensive medical examinations
- **Medication compliance**: Take prescribed medications consistently and as directed
- **Hydration habits**: Drink plenty of water daily (aim for 8-10 glasses)
- **Regular exercise**: Establish civilian fitness goals based on health rather than military standards
- **Enjoyable activities**: Find physical activities you enjoy rather than just effective exercises
- **Injury care**: Address service-related injuries with appropriate medical care
- **Self-care activities**: Engage in activities that promote mental and physical wellness
- **Sleep hygiene**: Maintain consistent sleep schedules and quality rest
- **Nutrition planning**: Eat balanced, nutritious meals that fuel your body properly
- **Stress management**: Use fitness as a stress management tool during civilian transition

Veteran Fitness Considerations:

- Adapt military fitness standards to civilian health and longevity goals
- Seek appropriate treatment and Veterans Affairs (VA) Compensation for service-related injuries
- Use fitness as a way to maintain military discipline in civilian life
- Find balance between fitness commitment and other life areas
- Take personal responsibility for health management, which the military previously coordinated
- Address both physical and mental health needs through a comprehensive wellness approach
- Don't neglect preventive care - early detection and treatment are crucial

5. Friendships - Building Civilian Relationships While Maintaining Military Bonds

Military friendships were built through shared sacrifice and a common mission. Civilian friendships develop differently and require new approaches while maintaining valuable military relationships. Social connections are essential for mental health, and both men and women need dedicated time with same-gender friends.

Friendships in Veteran Context:

- **Military friendship maintenance**: Keeping connections with military friends and mentors
- **Civilian relationship building**: Developing friendships with non-veterans
- **Social skill adaptation**: Learning civilian social customs and expectations
- **Trust development**: Building trust with civilians through different methods than the military
- **Community integration**: Becoming part of civilian social communities
- **Mental health support**: Recognizing friendships as essential for psychological well-being
- **Gender-specific bonding**: Participating in boys' night or girls' night activities

Friendship Development Strategies:

- **Regular military contact**: Maintain consistent communication with important military friends
- **Civilian group participation**: Join civilian groups based on interests and values
- **Patience practice**: Allow time for civilian relationship development
- **Experience sharing**: Share military experience appropriately to build understanding
- **Social integration**: Create opportunities for military and civilian friends to meet
- **Boys' night/Girls' night**: Participate in regular same-gender social activities for mental health
- **Relationship respect**: Honor boundaries and rules established by your significant other regarding social activities (drinking, smoking, gambling, fighting, late nights, etc.)
- **Mature behavior**: Keep foolishness to a minimum when socializing with friends and old military buddies

Veteran Friendship Considerations:

- Accept that civilian friendships develop differently from military bonds
- Don't expect immediate depth in civilian relationships
- Value both military and civilian friends for different qualities
- Use military friendship skills (loyalty, commitment) in civilian contexts
- Respect your partner's reasonable boundaries about social activities
- ***Avoid behaviors that could damage your primary relationship for the sake of friendships***
- Balance social time with friends and family responsibilities
- Choose friends who support your civilian goals and family commitments
- Healthy friendships should enhance, not threaten, your family relationships

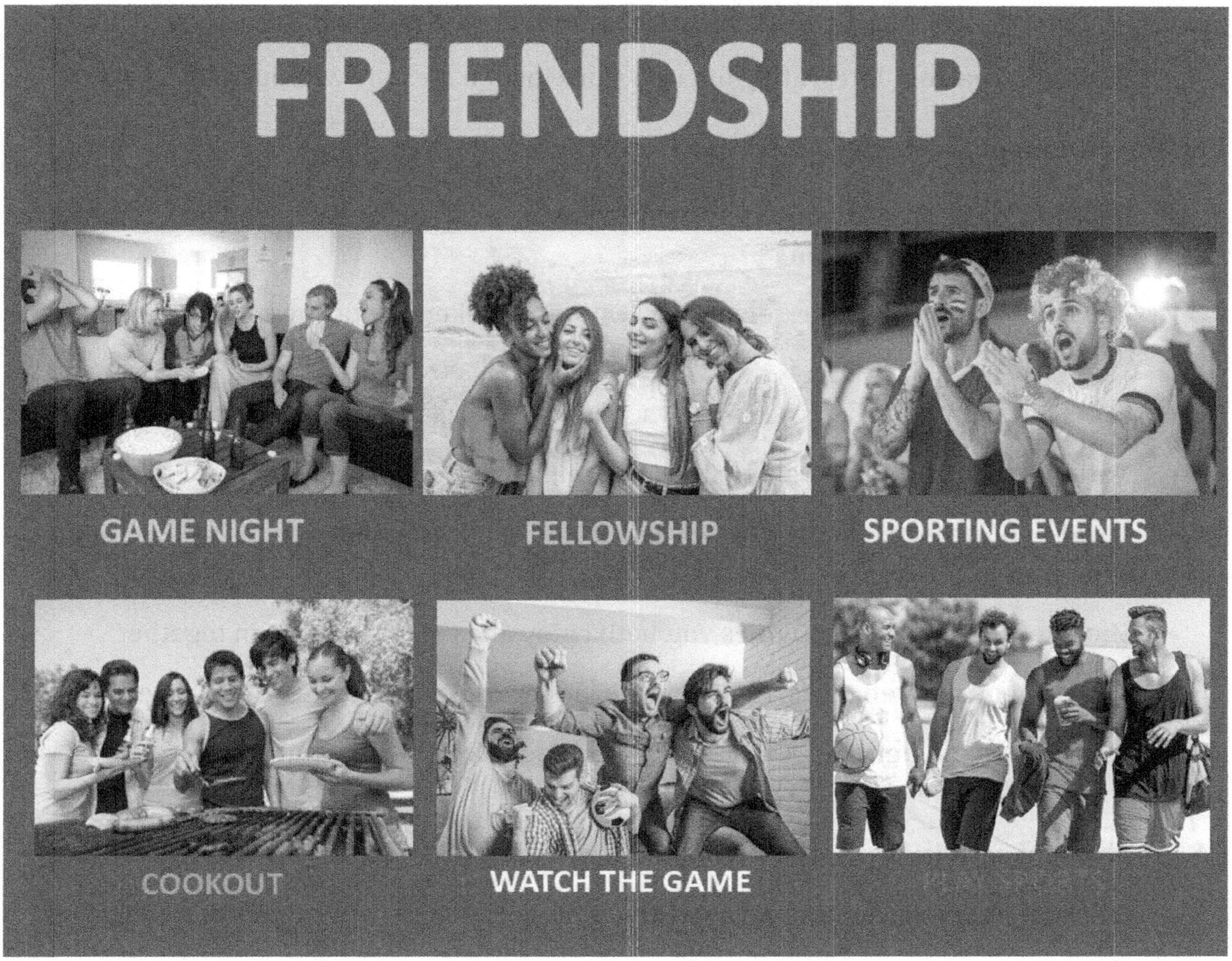

6. Fun - Connecting with Your Significant Other Through Shared Joy

Military service often requires sacrificing personal enjoyment for mission accomplishment. Civilian life provides an opportunity to rediscover fun and recreation, with particular emphasis on building connection and intimacy with your significant other through shared enjoyable experiences.

Fun in Veteran Context:

- **Joy rediscovery**: Learning to enjoy activities without a military purpose or mission
- **Partner connection**: Prioritizing fun activities that strengthen your relationship with your significant other
- **Recreational exploration**: Trying new activities and experiences together
- **Stress relief**: Using fun activities to manage transition and civilian stress as a couple
- **Relationship building**: Sharing enjoyable experiences to deepen intimacy and connection
- **Balance creation**: Integrating fun into a disciplined lifestyle without guilt

Fun Development Strategies with Your Partner:

- **Regular date nights**: Schedule consistent weekly or bi-weekly dates focused on enjoyment and connection
- **Vacation planning**: Take romantic vacations, weekend getaways, and extended trips together
- **Cruise adventures**: Explore cruise vacations for relaxation and shared experiences
- **Nature activities**: Take walks in the park, on the beach, or through scenic areas together
- **Local exploration**: Discover new restaurants, entertainment venues, and activities in your area
- **Shared hobbies**: Find activities you both enjoy and can participate in together
- **Spontaneous adventures**: Allow for unplanned fun activities and discoveries
- **Intimate experiences**: Create opportunities for meaningful conversations and emotional connection
- **Physical activities**: Engage in couples' fitness activities, dancing, or sports together
- **Cultural experiences**: Attend concerts, shows, museums, or festivals as a couple

Veteran Fun Considerations:

- Give yourself permission to enjoy activities that don't have clear purpose
- Prioritize your relationship with your significant other through dedicated fun time
- Experiment with different types of recreation to find what brings you both joy
- Use recreational activities to strengthen your romantic relationship
- Balance fun with other responsibilities without military-style guilt
- Remember that relationship investment through fun activities pays dividends in all areas of life
- Focus on creating positive memories that will sustain your relationship through challenges

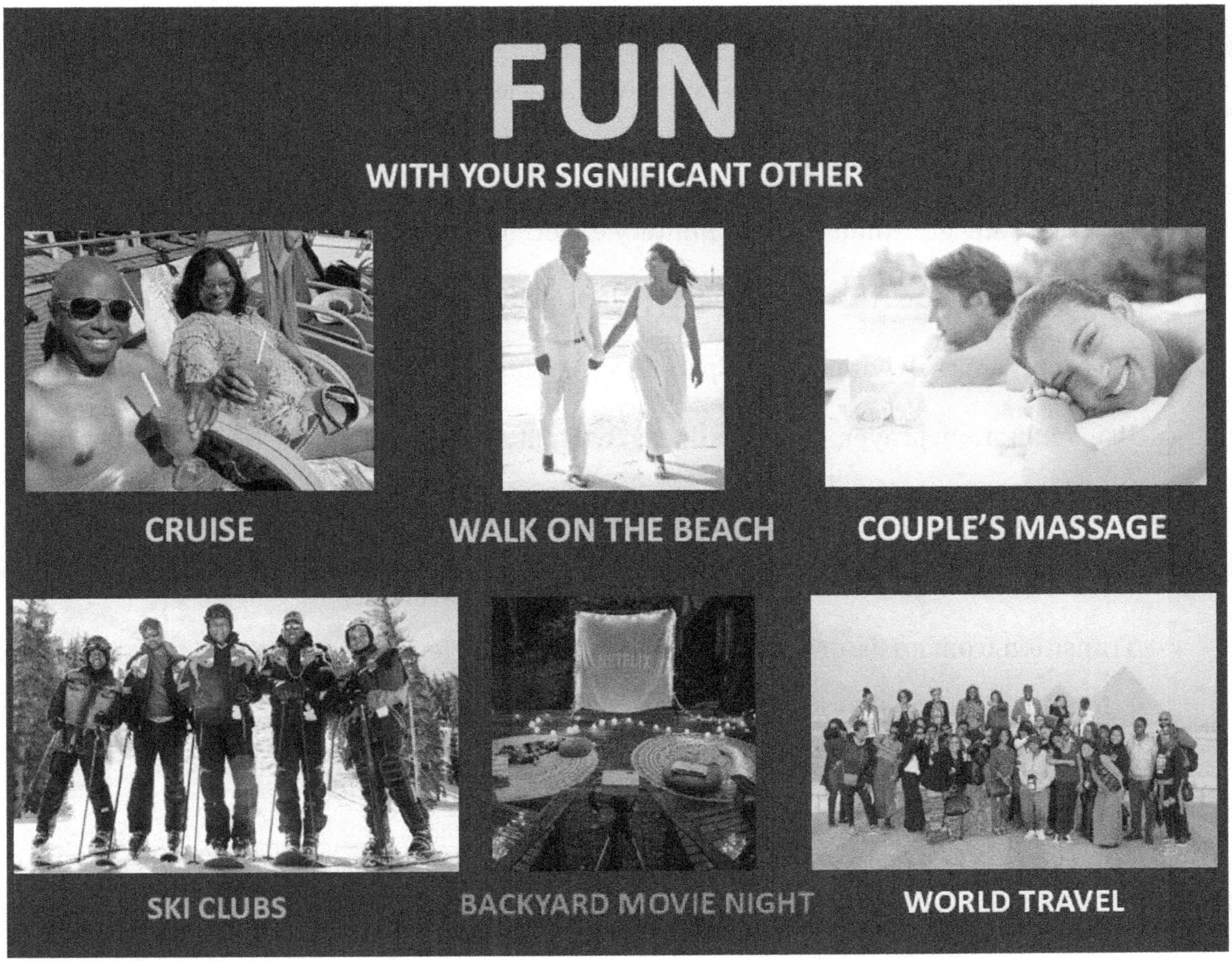

7. Frequency - Maintaining High Vibrational Energy

Frequency refers to your overall energy, attitude, and approach to life. Military service often created high-stress frequencies necessary for mission success. Civilian life allows for higher, more positive frequencies focused on joy, peace, and spiritual growth.

Frequency in Veteran Context:

- **Energy management**: Shifting from high-stress military frequency to balanced civilian energy
- **Attitude optimization**: Maintaining positive outlook during transition challenges
- **Spiritual elevation**: Connecting with higher purposes and meanings in civilian life
- **Emotional regulation**: Managing frequencies that serve civilian relationships and success
- **Vibration alignment**: Attracting positive experiences through high-frequency living

Frequency Development Strategies:

- Practice daily gratitude for both military experience and civilian opportunities
- Engage in activities that elevate your mood and energy
- Surround yourself with positive people and influences
- Use meditation, prayer, or mindfulness to maintain high frequency
- Focus on service to others as way to maintain purpose and meaning

Veteran Frequency Considerations:

- Transition from crisis-driven frequency to growth-oriented frequency
- Use military discipline to maintain consistent high-frequency practices
- Address negative frequencies from difficult military experiences
- Align civilian activities with frequencies that serve your highest good

FREQUENCY
LOVE, JOY, PEACE and SPIRITUAL ENLIGHTENMENT
HIGH FREQUENCY/VIBRATION
ATTRACTS POSITIVE EXPERIENCES
AND PEOPLE IN YOUR LIFE
GRATITUDE, SELF-CARE AND
UNCONDITIONAL LOVE
ARE MAJOR ASPECTS OF
RAISING YOUR FREQUENCY
TAKE 100% ACCOUNTABILITY
FOR YOUR FAMILY'S FREQUENCY

Implementing the F7 Method™ in Veteran Life

Success with Dr. Rod's F7 Neuro Mindset Method™ requires systematic implementation that fits veteran preferences for structure, measurement, and achievement.

Daily F7 Integration:

- **Morning F7 Review**: Brief assessment of each area during morning routine
- **Weekly F7 Planning**: Dedicated time for goal setting and progress review
- **Monthly F7 Evaluation**: Comprehensive assessment and adjustment of strategies
- **Quarterly F7 Visioning**: Big-picture review and long-term goal refinement

F7 Goal Setting Framework:

Each F7 area should have:

- **Short-term goals** (1-3 months)
- **Medium-term goals** (6-12 months)
- **Long-term goals** (1-5 years)
- **Daily practices** that support area development
- **Measurable outcomes** that track progress

F7 Balance Management:

- Recognize that different life seasons may emphasize different F7 areas
- Ensure no area is completely neglected for extended periods
- Use strong areas to support development in weaker areas
- Maintain minimum standards in all areas while excelling in priorities

Incorporating the Emotional Bucket Concept

Managing your Emotional Bucket is essential for the successful implementation of F7. When your bucket is full of transition stress and unprocessed military experiences, it becomes challenging to make progress in any F7 area.

Impact of a Full Emotional Bucket on F7 Development:

- **Reduced motivation** for positive activities across all areas
- **Difficulty maintaining balance** as stress overwhelms your systematic approach
- **Impatience with gradual progress** leading to abandonment of F7 practices
- **Negative frequency** that undermines all other F7 areas

Benefits of Managing Your Emotional Bucket:

- **Enhanced energy** for F7 area development and maintenance
- **Clear thinking** for effective goal setting and progress evaluation
- **Positive attitude** that supports high frequency and relationship building
- **Sustained motivation** for long-term F7 implementation

Exercise

F7 NEURO MINDSET METHOD™ ASSESSMENT AND DEVELOPMENT

Objective: Assess your current state in all seven F7 areas using Dr. Rod's method and develop comprehensive improvement strategies for balanced living.

Part 1: Current State Assessment

Rate each area from 1-10 (1=Needs Major Improvement, 10=Excellent):

Faith: __/10

- Current spiritual practices: ____
- Areas for improvement: ____

Family: __/10

- Current family relationship status: ______________________________________
- Areas for improvement: ____

Finances: __/10

- Current financial management: ____
- Areas for improvement: ____

Fitness: __/10

- Current physical activity level: ____
- Areas for improvement: ____

Friendships: __/10

- Current social connections: ____
- Areas for improvement: ____

Fun: __/10

- Current recreational activities: ____

- Areas for improvement: ____________________

Frequency: __/10

- Current energy/attitude level: ____________________
- Areas for improvement: ____________________

Part 2: F7 Goal Setting

For each area, set goals in three timeframes:

Faith Goals:

- 30-day goal: ____________________
- 6-month goal: ____________________
- 1-year goal: ____________________

Family Goals:

- 30-day goal: ____________________
- 6-month goal: ____________________
- 1-year goal: ____________________

Finance Goals:

- 30-day goal: ____________________
- 6-month goal: ____________________
- 1-year goal: ____________________

Fitness Goals:

- 30-day goal: ____________________
- 6-month goal: ____________________
- 1-year goal: ____________________

Friendship Goals:

- 30-day goal: ______________________________
- 6-month goal: ______________________________
- 1-year goal: ______________________________

Fun Goals:

- 30-day goal: ______________________________
- 6-month goal: ______________________________
- 1-year goal: ______________________________

Frequency Goals:

- 30-day goal: ______________________________
- 6-month goal: ______________________________
- 1-year goal: ______________________________

Part 3: Daily F7 Practices

Identify specific daily practices for each area:

- ☐ **Faith:** Daily 10-minute spiritual practice: ______________
- ☐ **Family:** Daily family connection activity: ______________
- ☐ **Finances:** Daily financial management task: ______________
- ☐ **Fitness:** Daily physical activity: ______________
- ☐ **Friendships:** Daily social connection: ______________
- ☐ **Fun:** Daily enjoyable activity: ______________
- ☐ **Frequency:** Daily frequency-raising practice: ______________

Part 4: Weekly F7 Schedule

Plan weekly time allocation for focused F7 development:

- ☐ **Monday Focus:** ______ (Choose F7 area)
- ☐ **Tuesday Focus:** ______ (Choose F7 area)
- ☐ **Wednesday Focus:** ______ (Choose F7 area)
- ☐ **Thursday Focus:** ______ (Choose F7 area)
- ☐ **Friday Focus:** ______ (Choose F7 area)
- ☐ **Saturday Focus:** ______ (Choose F7 area)
- ☐ **Sunday Focus:** ______ (Choose F7 area)

Part 5: F7 Integration Strategy

Identify ways to integrate multiple F7 areas:

- ☐ **Faith + Family:** ______________________________
- ☐ **Fitness + Friendships:** ______________________________
- ☐ **Fun + Family:** ______________________________
- ☐ **Finances + Fitness:** ______________________________
- ☐ **Frequency + Faith:** ______________________________

Self-Reflective Questions

1. Which of the seven F7 areas do I most neglect, and how has this affected my overall well-being during civilian transition?

2. How can I adapt the structured approach of military life to systematically develop all F7 areas in civilian contexts?

3. Which F7 areas were strongest during my military service, and how can I leverage those strengths to improve weaker areas?

4. How does the state of my Emotional Bucket impact my ability to maintain balance across all seven F7 areas?

5. What specific military values and skills can I apply to create discipline and consistency in F7 development?

Affirmations

1. I am committed to creating balance in all areas of my life through systematic F7 development.

2. My military discipline and goal-oriented approach serve me well in building a comprehensive civilian lifestyle.

3. Each day I grow stronger in faith, family, finances, fitness, friendships, fun, and frequency.

Actionable Steps

1. **Complete F7 Assessment**: Use the assessment tool to honestly evaluate your current state in all seven areas and identify priorities for improvement.

2. **Implement Daily F7 Practices**: Establish specific daily practices for each F7 area and integrate them into your morning or evening routine.

3. **Create Weekly F7 Focus**: Dedicate specific time each week to focused development in each F7 area through the weekly schedule approach.

4. **Track F7 Progress**: Maintain weekly tracking of your F7 goals and adjust strategies based on what works best for your lifestyle.

5. **Integrate F7 Areas**: Look for opportunities to combine F7 development activities (like exercising with friends or having family financial discussions) for maximum efficiency.

Creating Your Balanced Veteran Lifestyle

Dr. Roderick Cunningham's F7 Neuro Mindset Method™ provides veterans with the comprehensive structure needed to thrive in civilian life. By systematically developing all seven areas, you create the balanced foundation that military service once provided while embracing the opportunities that civilian life offers.

Remember that balance doesn't mean equal attention to all areas at all times. Different life seasons may require emphasis on different F7 areas. The key is maintaining awareness of all seven areas and ensuring that none are completely neglected for extended periods.

Your military experience provides significant advantages in F7 Method™ implementation: discipline for daily practices, goal-oriented mindset for systematic improvement, leadership skills for family and social situations, and service orientation for maintaining high frequency. Use these military strengths to build the civilian life you deserve.

The F7 Neuro Mindset Method™ is not just about personal development - it's about honoring your military service by building a civilian life worthy of the sacrifice and discipline that brought you to this point. By excelling in all seven areas, you demonstrate that veterans can not only succeed in civilian life but can model balanced, purposeful living for others.

Evidence-Based Research: Research on holistic wellness approaches shows that individuals who systematically address multiple life domains report significantly higher life satisfaction and resilience compared to those who focus on single areas (Seligman, 2011). Studies specific to veterans indicate that comprehensive lifestyle approaches that integrate physical, social, spiritual, and financial wellness lead to better transition outcomes and reduced symptoms of depression and anxiety (Beder, 2012). Additionally, veterans who maintain structured personal development approaches similar to military training report higher civilian career satisfaction and community integration (Romaniuk & Romaniuk, 2018).

Notes

Goal Setting for Success

Chapter 15

Goal setting is an essential skill that acts as a catalyst for success in both personal and professional spheres. Effective goal setting provides direction, boosts motivation, and facilitates significant achievements. This chapter delves into the importance of setting SMART goals, offering a structured approach to define, prioritize, and track these goals to maximize potential and achieve substantial outcomes.

The Importance of SMART Goals

SMART goals offer a proven framework for setting clear and achievable objectives. The acronym SMART stands for:

- **Specific**: Goals should be well-defined to direct focus and effort.
- **Measurable**: Progress towards goals should be quantifiable.
- **Achievable**: Goals need to be realistic and attainable within available resources.
- **Relevant**: Goals should align with broader life objectives and have a clear purpose.
- **Time-bound**: Assigning deadlines provides motivation and a sense of urgency.

Employing SMART goals ensures objectives are framed to harness effort effectively towards meaningful results, helping to avoid the pitfalls of vague or unrealistic goals that can lead to frustration and decreased productivity.

Step-by-Step Guide on How to Set and Prioritize Goals

1. **Define Your Objectives:**
    - Begin by identifying what you want to achieve across various life areas such as faith, family, finances (to include education and career), fitness, fun, and friendships.
2. **Apply the SMART Criteria:**
    - Refine each goal to ensure it adheres to all elements of the SMART framework. For example, transform a general goal like "I want to improve my coding skills" into a SMART goal: "I aim to complete a Python programming course by the end of Q3 to enhance my coding proficiency."
3. **Prioritize Your Goals:**
    - Evaluate and order your goals by their impact on your life and the urgency with which they need to be achieved.

4. **Create Action Plans:**
    - Decompose each goal into actionable steps, assigning deadlines and identifying necessary resources for each milestone. This approach makes goals less overwhelming and clarifies the path to achievement.
5. **Regular Review and Adjustment:**
    - Continuously assess your progress and adapt your strategies as needed. This flexibility allows you to stay aligned with your goals despite any changes in circumstances.

Worksheet for Goal Setting and Tracking Progress To effectively monitor your goals, use a structured worksheet that includes:

- **Goal Description**: Clearly articulate each goal, ensuring it meets the SMART criteria.
- **Motivation**: Document the reasons each goal is important to you, which will serve as motivation.
- **Milestones**: Outline key milestones with deadlines to structure your approach to achieving your goals.
- **Progress Logs**: Maintain a section to note updates, challenges, and adjustments.
- **Outcome Review**: After reaching a deadline, evaluate the results: Was the goal met? What were the contributing factors or obstacles? What lessons can be drawn?

Goal setting is a strategic process that, when executed using the SMART framework, can lead to substantial and tangible success. By following the outlined steps and utilizing the tracking worksheet, you are equipped to set goals that are clear, achievable, and motivating, aligning with your overall objectives in life. This methodical approach ensures you remain focused and efficient on your journey towards personal and professional fulfillment.

Incorporating the Emotional Bucket Analogy in Goal Setting

The Emotional Bucket is a useful metaphor for understanding how unmanaged stress and emotions can impact goal-setting abilities. When the bucket is full, it represents an overload of unmanaged stress and emotions, which can:

- **Decrease Clarity and Focus**: Emotional clutter makes it challenging to prioritize goals and see the necessary steps for achievement.
- **Reduce Motivation**: High stress and emotional turmoil can drain the energy required to pursue even well-defined goals.
- **Impair Decision Making**: Emotional overload can cloud judgment, leading to poorly set goals that may not truly align with one's aspirations.

Positive Impact of Managing the Emotional Bucket

- **Enhanced Clarity and Concentration**: Clearing the Emotional Bucket allows for more focused and effective goal setting.
- **Boosted Motivation and Energy**: Reducing emotional clutter reinvigorates motivation and energy, essential for pursuing goals.
- **Improved Decision Making**: Better emotional regulation enhances cognitive functions, aiding in more rational and informed goal setting.

Integrating Emotional Management in Goal Setting

- **Mindfulness and Reflection**: Engage in mindfulness practices before setting goals to achieve the mental clarity required for effective planning.
- **Emotional Check-Ins**: Regularly assess your emotional state to ensure it supports, rather than hinders, your goal-setting sessions.
- **Incorporate Emotional Wellness Goals**: Include objectives for emotional health to maintain an effectively managed Emotional Bucket, thereby supporting overall success.

Developing a Vision Board with the F7 Neuro Mindset Method: The F7 Neuro Mindset Method focuses on the seven facets of a fulfilling life: Faith (***spirit***), Family, Finances, Fitness (***body***), Friendship, Fun, and Frequency (***mind***). To create a vision board using this method, gather images, quotes, and symbols representing your aspirations in these seven areas. Arrange them on a board in a way that feels inspiring. Your vision board is a visual representation of your future self. It trains your mind to believe that you have already achieved what's on the board. Place your vision board somewhere you will see it daily to keep your goals visually and mentally in focus. You can also create your vision board in PowerPoint or Canva.

Watch these YouTube Videos weekly, and they WILL change your life:

Watch First | The Secret | Beyond the Secret

"SAMPLE" VISION BOARD

FAITH

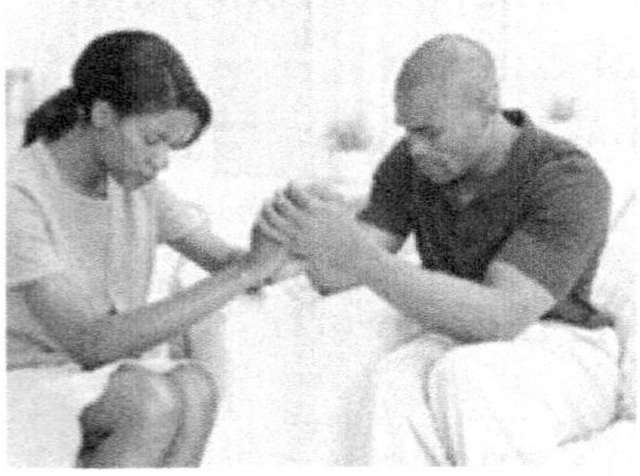

Gain Spiritual Enlightenment

FAMILY

Dream Wedding

FITNESS

FINANCES

Bachelor's Degree by age 22
Master's Degree by age 24
MD, J.D., or PhD by age 30

$200K per year
in income by age 30

Dream Home

Dream Car

FRIENDSHIPS

FUN

FREQUENCY

Love, Joy, and Peace

Exercise

GOAL SETTING

Setting Goals

Goals are the benchmarks that guide your journey towards personal and professional success. Start by defining long-term goals for each of the F7 Neuro categories, envisioning where you want to be in 5, 10, or 20 years. Break these down into short-term goals that you can achieve within the next year. These should be specific, measurable, achievable, relevant, and time-bound (SMART). Regularly review and adjust these goals to reflect your growth and any changes in your life direction.

Examples of short-term goals:

	FAITH
	I will read the Bible at least 5 times weekly by Jan 1, 2027
	I will volunteer as an usher at my church by Jun 1, 2027
	FAMILY
	I will eat dinner at the table consistently with my wife and children by Mar 1, 2027
	I will do fun activities with my family every weekend by Jan 1, 2027
	I will start reconnecting with my younger brother and forgive him by Jan 1, 2027
	FINANCES
	I will graduate on time with my bachelor's/master's degree by Jun 15, 2027
	I will pay myself 30% of my paycheck before I pay anyone else (invest 10%, save 10%, and give 10% to church/charity) by Jun 1, 2027
	I will use my phone contacts and LinkedIn to reach out to 20 people per week to develop professional relationships. These relationships will lead to me landing my perfect career that pays $120,000 per year by Aug 15, 2027
	I will earn $15,000 per month through my business ventures by May 1, 2028

	FITNESS
	I will run 2-miles under 18 minutes by Jun 15, 2027
	I will increase my water intake from two glasses per day to four by Feb 1, 2027
	FRIENDSHIP
	I will attend more events with my friends if the events are safe for all involved by Jun 20, 2028.
	FUN
	I will take weekly walks with my significant other and always give them the best of me by Aug 25, 2027,
	We will take a romantic vacation for 14 days a year by Jun 20, 2028
	FREQUENCY
	I will review my vision board daily, be grateful for all that I have, visualize (my life in 20 years) each day, and meditate day and night by May 1, 2027
	I will love myself and others unconditionally, find peace, joy, and spiritual enlightenment by May 1, 2027
	I will practice self-care daily by June 1, 2028

Long-Term Goals

You plan to achieve long-term goals over the next 1 to 20 years. With long-term goals, you should block out 30 minutes each morning, close your eyes, and visualize your future at the age of 50. After that visualization exercise, you can work on the steps today to get you to your goal at the age of 50. For example, if you want to be a high school principal, then you must first have a master's degree and you must have spent years as a teacher. In order to be a teacher, you must have a bachelor's degree. To get a bachelor's degree, you must go to college. To go to college, you must get excellent grades. To get excellent grades, you must minimize absences, study, and respect the education system. The things that you do today, will prepare you to become a high school principal at 50 years old.

Examples of long-term goals:

	FAITH
	I will read the entire Bible by age 33
	FAMILY
	I will get married and have two children by age 35
	FINANCES
	I will become a millionaire by age 34
	I will be debt-free by age 29
	FITNESS
	I will run 4 miles per day by age 29
	I will work out at least 5x per week by age 27
	FRIENDSHIP
	I will get NFL season tickets and purchase two more for my friends by age 30
	FUN
	I will take a 30-day vacation annually with my spouse
	I will take my wife on a date at least twice per month
	FREQUENCY
	I will continue to review my vision board daily, be grateful for all that I have, visualize my life in 20 years each day, and meditate day and night
	I will love myself and others unconditionally, have peace, joy, and spiritual enlightenment

YOUR GOALS

SHORT-TERM GOALS

	FAITH	Date or Age
	FAMILY	
	FINANCE	
	FITNESS	
	FRIENDSHIPS	
	FUN	
	FREQUENCY	

LONG-TERM GOALS

	FAITH	Date or Age
	FAMILY	
	FINANCE	
	FITNESS	
	FRIENDSHIPS	
	FUN	
	FREQUENCY	

Self-Reflective Questions

1. How often do I feel overwhelmed by emotions when planning my future?
2. Have I ever set a goal while feeling stressed or emotional, and later found it wasn't truly what I wanted?
3. What emotional management techniques can I integrate into my goal-setting process?
4. How does my emotional state affect my motivation and focus on achieving my goals?
5. Can I identify a time when managing my emotions helped me achieve a goal more effectively?

Affirmations

1. I maintain clarity and focus by managing my Emotional Bucket effectively.
2. Every goal I set is aligned with my true desires and capabilities.
3. I am motivated and energized to achieve my best.
4. I make informed and rational decisions that reflect my personal and professional aspirations.
5. I am committed to achieving emotional wellness as part of my overall success.

Actionable Steps

1. **Schedule Regular Emotional Clean-ups**: Set aside time each week to address and process emotional stressors, keeping the Emotional Bucket from overflowing.
2. **Integrate Mindfulness Into Daily Routine**: Practice mindfulness or meditation daily to enhance mental clarity, which will aid in setting more focused and achievable goals.
3. **Set Emotional Wellness Goals**: Explicitly include emotional wellness as a category in your goal-setting to maintain balance.
4. **Use a Goal-Setting Worksheet**: Apply a structured approach to track not only your goals but also your emotional wellness over time.
5. **Evaluate Goals in a Low-Stress Environment**: Choose a calm, relaxed environment for goal review and planning sessions to avoid decision-making under stress.

Evidence-Based Research

Research supports the integration of emotional wellness into goal setting. Studies indicate that stress management can significantly enhance cognitive functions related to planning and execution (McEwen, B.S., 2007). Furthermore, mindfulness practices have been shown to improve focus and decision-making abilities, key components of effective goal setting (Kabat-Zinn, J., 1994). By managing emotional load, individuals can enhance their capacity to set and achieve meaningful and ambitious goals.

Fitness - Mental and Physical Readiness

Chapter

16

Physical fitness was a cornerstone of military life, serving not only as a requirement for service but as a foundation for mental toughness, discipline, and mission readiness. As you transition to civilian life, fitness continues to play a crucial role, but with different objectives and approaches. This chapter explores how to adapt your military fitness mindset to civilian wellness goals while maintaining the mental and physical readiness that served you well in service and will continue to benefit you in civilian life.

Understanding the Transition from Military to Civilian Fitness

Military fitness served specific purposes: meeting physical standards, maintaining unit readiness, building mental toughness, and ensuring mission capability. Civilian fitness serves different but equally important purposes: long-term health, stress management, quality of life, injury prevention, and mental wellness.

Military Fitness Characteristics:

- **Standards-based**: Meeting specific requirements and benchmarks
- **Mission-focused**: Physical readiness for operational demands
- **Unit-oriented**: Group training and competitive elements
- **Time-constrained**: Mandatory PT sessions with limited scheduling flexibility
- **Youth-advantaged**: Designed for a generally young, healthy population
- **Performance-driven**: Emphasis on speed, strength, and endurance metrics
- **Injury-tolerant**: "Push through pain" mentality is common

Civilian Fitness Characteristics:

- **Health-focused**: Long-term wellness and disease prevention
- **Lifestyle-integrated**: Flexible scheduling around work and family
- **Individually-tailored**: Personal goals and preferences drive activity choices
- **Sustainable**: Emphasis on activities that can be maintained long-term
- **Injury-preventive**: Focus on avoiding injury and managing existing conditions
- **Holistic**: Integration of physical, mental, and emotional wellness
- **Quality-oriented**: Emphasis on how exercise makes you feel and function

The Mental Health Connection

For veterans, fitness serves as a powerful tool for mental health management during civilian transition. ***The discipline, routine, and endorphin release from exercise can help manage depression, anxiety, PTSD symptoms, and transition stress.***

When I'm stressed, I work out. When I'm angry, I work out. When I'm anxious, I work out. When I'm grieving, I work out. When I feel rejection, I work out. Daily exercise will help manage your emotional health and improve your mood and frequency. This is called self-care.

Mental Health Benefits of Fitness for Veterans:

- **Stress reduction**: Physical activity naturally reduces cortisol and stress hormones
- **Mood elevation**: Exercise triggers endorphin release, improving mood and outlook
- **Sleep improvement**: Regular activity promotes better sleep quality and duration
- **Anxiety management**: Physical exertion helps process anxiety and nervous energy
- **Depression combat**: Exercise serves as a natural antidepressant through neurochemical changes
- **PTSD symptom relief**: Physical activity can help process trauma and reduce hypervigilance
- **Routine establishment**: Fitness provides structure during uncertain transition periods
- **Self-efficacy building**: Physical achievements boost confidence in other life areas

Veteran-Specific Mental Health Considerations:

- Use fitness as a healthy coping mechanism instead of junk food, video games, sexual addiction, substances, or isolation
- Recognize that military fitness identity may need adjustment, but doesn't disappear
- Address service-related injuries that may affect exercise choices
- Find fitness communities that understand and respect a military background
- Use physical activity to process difficult military experiences in healthy ways

Building Your Civilian Fitness Foundation

Creating a sustainable civilian fitness routine requires adapting military discipline to civilian realities while maintaining the mental and physical benefits you gained from military fitness.

Civilian Fitness Planning Principles:

1. **Assess Your Current State:**
   - Document service-related injuries and physical limitations
   - Evaluate current fitness level honestly without military comparison
   - Identify activities you enjoy versus those you feel obligated to do
   - Consider time constraints from work and family responsibilities
2. **Set Realistic Goals:**
   - Focus on health and wellness rather than military standards
   - Establish both short-term and long-term objectives
   - Include both physical and mental health goals
   - Make goals specific, measurable, and achievable
3. **Choose Sustainable Activities:**
   - Select exercises you can realistically maintain long-term
   - Consider activities that fit your civilian schedule and budget
   - Include variety to prevent boredom and overuse injuries
   - Factor in seasonal and weather considerations
4. **Address Injuries and Limitations:**
   - Seek proper medical evaluation for service-related conditions
   - Work with healthcare providers to develop safe exercise plans
   - Learn modifications that allow continued activity despite limitations
   - Consider physical therapy or rehabilitation as needed

Comprehensive Wellness Approach

Civilian fitness should encompass more than just physical exercise. A comprehensive approach includes preventive healthcare, nutrition, mental wellness, and lifestyle factors that support overall health.

Physical Health Components:

Cardiovascular Fitness:

- Activities that strengthen the heart and improve circulation
- Options: walking, jogging, cycling, swimming, dancing, hiking
- Goal: 150 minutes moderate activity or 75 minutes vigorous activity weekly
- Benefits: heart health, endurance, mood improvement, stress reduction

Strength Training:

- Maintaining muscle mass and bone density as you age
- Options: weightlifting, resistance bands, bodyweight exercises, functional movements
- Goal: 2-3 strength sessions per week targeting all major muscle groups
- Benefits: injury prevention, metabolism support, functional daily activities

Flexibility and Mobility:

- Maintaining range of motion and preventing stiffness
- Options: yoga, stretching routines, tai chi, foam rolling
- Goal: Daily flexibility work or dedicated sessions 3-4 times weekly
- Benefits: injury prevention, pain reduction, stress relief, better movement quality

Balance and Coordination:

- Important for injury prevention, especially as you age
- Options: balance exercises, agility drills, martial arts, stability work
- Goal: Include balance challenges in regular routine
- Benefits: fall prevention, athletic performance, brain health

Preventive Healthcare Integration:

Annual Health Physicals:

- Schedule comprehensive medical examinations yearly
- Include cancer screenings appropriate for your age and risk factors
- Monitor blood pressure, cholesterol, blood sugar, and other vital markers

- Discuss exercise plans with healthcare providers

Medication Compliance:

- Take prescribed medications consistently and as directed
- Understand how medications may affect exercise performance or safety
- Communicate with healthcare providers about exercise and medication interactions
- Monitor for side effects that might impact physical activity

Hydration and Nutrition:

- Drink plenty of water daily (aim for 8-10 glasses minimum)
- Eat balanced, nutritious meals that fuel your physical activity
- Consider nutritional needs that may change with age or activity level
- Limit alcohol and substances that can interfere with fitness goals

Sleep Hygiene:

- Maintain consistent sleep schedules (7-9 hours nightly)
- Create a sleep environment conducive to quality rest
- Understand how exercise timing affects sleep quality
- Address sleep disorders that may impact recovery and performance

Self-Care Activities:

- Integrate stress-reducing activities into your routine
- Include activities like meditation, massage, hobbies, or relaxation techniques
- Recognize self-care as an essential component of fitness, not a luxury
- Balance physical activity with recovery and restoration

Adapting Military Fitness Discipline to Civilian Life

Your military fitness experience provides significant advantages that can be adapted for civilian success:

Military Advantages to Leverage:

Discipline and Routine:

- Use military discipline to maintain consistent exercise habits
- Adapt PT schedule concepts to civilian morning or evening routines
- Apply military time management skills to prioritize fitness
- Use goal-setting and measurement techniques from military training

Mental Toughness:

- Apply military mental resilience to overcome exercise obstacles
- Use a military "can-do" attitude to push through difficult workouts
- Leverage experience with discomfort to challenge yourself appropriately
- Draw on military teamwork experience in group fitness settings

Knowledge and Experience:

- Apply understanding of progressive training and periodization
- Use knowledge of proper form and exercise safety
- Leverage experience with different types of physical challenges
- Adapt military fitness test concepts to personal benchmarking

Adaptations Needed for Civilian Context:

Flexibility and Balance:

- Allow more flexibility in scheduling around family and work demands
- Balance fitness goals with other life priorities
- Adapt intensity based on recovery needs and age considerations
- Include rest and recovery as essential components

Injury Prevention Focus:

- Shift from "push through pain" to "listen to your body"
- Prioritize form and safety over performance metrics
- Address existing injuries before they become chronic problems
- Include warm-up, cool-down, and recovery activities

Sustainable Intensity:

- Adjust intensity for long-term health rather than short-term performance
- Include variety to prevent overuse and maintain motivation
- Focus on consistency over extremes
- Consider age-appropriate modifications

Managing Service-Related Injuries and Limitations

Many veterans have service-related injuries that require special consideration in civilian fitness planning. These limitations don't eliminate fitness possibilities but require thoughtful adaptation.

Common Service-Related Conditions:

Joint Issues (knees, back, shoulders):

- Focus on low-impact activities like swimming, cycling, or elliptical
- Include strength training to support affected joints
- Use proper form and avoid aggravating movements
- Consider water-based exercises for joint-friendly activity

Traumatic Brain Injury (TBI):

- Start slowly and monitor for symptoms during exercise
- Avoid activities with high concussion risk
- Include balance and coordination exercises as appropriate
- Work with healthcare providers to develop safe exercise plans

PTSD and Mental Health:

- Use exercise as a healthy coping mechanism for symptoms
- Find fitness environments that feel safe and comfortable
- Consider individual activities if group settings are triggering
- Include mindfulness-based activities like yoga or walking meditation

Chronic Pain:

- Work with healthcare providers to develop appropriate activity levels
- Focus on activities that reduce rather than increase pain
- Include gentle movement and stretching as pain allows
- Consider physical therapy or specialized programs

Hearing Loss:

- Adapt group fitness participation as needed for communication
- Use visual cues and written instructions when necessary
- Find fitness communities that are understanding of hearing limitations
- Consider individual activities if group settings are challenging

Building Your Veteran Fitness Community

Finding the right fitness community can provide motivation, support, and understanding while you adapt to civilian fitness goals.

Veteran Fitness Communities:

Veteran-Specific Programs:

- Look for gyms or programs specifically designed for veterans
- Join veteran running clubs, cycling groups, or sports leagues
- Participate in veteran fitness challenges or competitions
- Connect with adaptive sports programs if dealing with injuries

Inclusive Civilian Communities:

- Find gyms and fitness centers that welcome veterans
- Join community sports leagues or recreational groups
- Participate in charity fitness events that support veteran causes
- Look for fitness professionals who understand the military

Online and Virtual Communities:

- Join online veteran fitness groups for support and motivation
- Use fitness apps that connect you with other veterans
- Participate in virtual challenges with veteran friends
- Share fitness goals and progress with veteran support networks

Incorporating the Emotional Bucket Concept

Managing your Emotional Bucket is crucial for maintaining consistent fitness habits and achieving mental health benefits from physical activity. When your bucket is full of transition stress and unprocessed emotions from past grief, abandonment, rejection, and trauma, it becomes difficult to maintain exercise motivation and routine.

Impact of a Full Emotional Bucket on Fitness:

- **Reduced motivation** for physical activity when overwhelmed by stress
- **Inconsistent routines,** like emotional chaos, disrupt exercise habits
- **Injury risk increases** when stress affects focus and form
- **Negative self-talk** that undermines fitness confidence and goals

Benefits of Managing Your Emotional Bucket:

- **Enhanced motivation** for consistent physical activity
- **Better stress relief** as exercise helps process emotions effectively
- **Improved safety** with better focus and form during exercise
- **Positive momentum** where fitness success builds emotional resilience

Exercise

VETERAN FITNESS ASSESSMENT AND PLANNING

Objective: Develop a comprehensive civilian fitness plan that addresses physical health, mental wellness, and service-related considerations.

Part 1: Current State Assessment

Physical Health Evaluation:

- Current fitness level (1-10): ______
- Service-related injuries or limitations: ____
- Current physical activities: ____
- Time available for fitness weekly: ______ hours
- Fitness equipment access: ____

Mental Health and Motivation:

- How often do you use exercise for stress relief? __________
- Current stress level (1-10): ______
- Mental health challenges affecting fitness: ____
- Motivation level for civilian fitness (1-10): ______

Healthcare Status:

- Date of last physical exam: ____
- Current medications affecting exercise: ____
- Healthcare provider supporting fitness goals: _______________________________________
- Outstanding health concerns: ____

Part 2: Fitness Goal Setting

Short-term Goals (1-3 months):

- Physical goal: ____
- Mental health goal: ____
- Habit development goal: ____

Medium-term Goals (3-6 months):

- Fitness achievement goal: ______________________________
- Health improvement goal: ______________________________
- Community involvement goal: ______________________________

Long-term Goals (6-12 months):

- Overall wellness vision: ______________________________
- Injury management goal: ______________________________
- Lifestyle integration goal: ______________________________

Part 3: Comprehensive Fitness Plan

Weekly Exercise Schedule:

Monday: ____________________ Duration: ________

Tuesday: ____________________ Duration: ________

Wednesday: ____________________ Duration: ________

Thursday: ____________________ Duration: ________

Friday: ____________________ Duration: ________

Saturday: ____________________ Duration: ________

Sunday: ____________________ Duration: ________

Healthcare Integration:

- Annual physical scheduled: ______________________________
- Medication review planned: ______________________________
- Injury management plan: ______________________________

Nutrition and Hydration Plan:

- Daily water goal: ______________________________
- Nutrition support for fitness: ______________________________

- Meal timing around exercise: ____

Part 4: Adaptation Strategies

Service-Related Considerations:

- Injury modifications needed: ____
- Activities to avoid: ____
- Adaptive equipment needed: ____

Mental Health Support:

- Exercise for stress relief plan: ____
- Fitness community involvement: ____
- Professional support if needed: ____

Part 5: Progress Tracking

Weekly Check-in Questions:

- Did I meet my exercise goals this week?
- How did exercise affect my mood and stress levels?
- What adjustments do I need to make next week?
- Am I maintaining balance with other life priorities?

Monthly Evaluation:

- Progress toward fitness goals: ____
- Changes in physical or mental health: ____________________________________
- Adjustments needed to plan: ____

Self-Reflective Questions

1. How has my relationship with fitness changed since leaving the military, and what adjustments do I need to make for civilian success?

2. What service-related injuries or limitations do I need to address to maintain long-term physical activity and health?

3. How effectively am I using exercise as a tool for managing transition stress and maintaining mental health?

4. What prevents me from maintaining consistent fitness habits in civilian life, and how can I overcome these obstacles?

5. How does the state of my Emotional Bucket affect my motivation and consistency with physical activity and self-care?

Affirmations

1. I maintain physical and mental fitness as a foundation for successful civilian life.
2. My military fitness experience provides valuable discipline and knowledge that I adapt for civilian wellness.
3. I prioritize my health through consistent exercise, proper nutrition, preventive care, and stress management.

Actionable Steps

1. **Complete Comprehensive Health Assessment**: Schedule annual physical exam and work with healthcare providers to address any service-related conditions affecting fitness.
2. **Develop a Realistic Fitness Plan:** Create a weekly exercise schedule that includes cardiovascular, strength, flexibility, and balance components appropriate for your current condition and goals.
3. **Establish Daily Wellness Habits**: Implement consistent hydration (8-10 glasses of water daily), take medications as prescribed, and include stress management activities in your routine.
4. **Join Fitness Community**: Connect with veteran fitness groups or civilian communities that support your goals and understand your background.
5. **Track Progress and Adjust**: Maintain weekly fitness logs and monthly evaluations to ensure your plan remains effective and sustainable for long-term health.

Maintaining Readiness for Civilian Life

Physical and mental fitness remains as essential in civilian life as it was in military service, though the objectives and methods may differ. Your military experience provides a strong foundation of discipline, knowledge, and mental toughness that serves you well in civilian fitness pursuits.

The key is adapting these strengths to civilian realities: longer-term health goals, family and work responsibilities, service-related injuries, and the need for sustainable rather than performance-driven fitness. By maintaining physical readiness, you preserve not only your health but also the confidence, discipline, and resilience that served you well in military service.

Remember that civilian fitness is not about meeting external standards but about maintaining the physical and mental capabilities that allow you to thrive in all aspects of civilian life. Your fitness routine should support your career, family relationships, and personal goals while providing the stress relief and mental health benefits essential for a successful transition.

Your commitment to fitness demonstrates the same dedication to excellence that defined your military service. By maintaining physical and mental readiness, you honor your military experience while building the foundation for long-term civilian success and well-being.

Evidence-Based Research: Studies show that veterans who maintain regular physical activity report significantly better mental health outcomes during civilian transition, including reduced symptoms of depression and anxiety (Rosenbaum et al., 2014). Research also indicates that exercise serves as an effective intervention for PTSD symptoms in veteran populations, with structured fitness programs showing measurable improvements in sleep quality, mood regulation, and overall life satisfaction (Whitworth et al., 2016). Additionally, veterans who adapt military fitness discipline to civilian wellness goals demonstrate better long-term health outcomes and successful transition adjustment compared to those who abandon fitness routines post-service (Caddick & Smith, 2018).

Notes

Discovering Your Dream Career – Post Service

Chapter 17

"A career is something that you train for and prepare for and plan on doing for a long time."

- Sonia Sotomayor

One of the most significant challenges veterans face is translating their military experience into a fulfilling civilian career that provides both financial security and personal satisfaction. Military service may have introduced you to skills and interests you never knew you had, or it may have confirmed natural talents and preferences. This chapter helps you identify your ideal career path by understanding your personality type, leveraging your military experience, and making strategic decisions about your professional future.

The Veteran Career Discovery Challenge

Unlike traditional college students who often explore careers during their educational years, veterans enter civilian career planning with significant life and work experience. This can be both an advantage and a challenge. Your military experience provides valuable skills and maturity, but it may also create uncertainty about how to apply these skills in civilian contexts or what careers will provide the same sense of purpose and mission you found in service.

Unique Veteran Career Considerations:

- **Purpose translation:** Finding civilian work that provides the same sense of mission and meaning as military service
- **Skills translation:** Understanding how military experience applies to civilian career opportunities
- **Identity integration:** Developing a professional identity that honors military service while embracing civilian opportunities
- **Compensation expectations:** Balancing financial needs with career satisfaction and growth potential
- **Geographic flexibility:** Considering how military mobility experience affects location preferences and limitations
- **Family impact:** Making career decisions that support family stability and spouse/children's needs
- **Time investment:** Evaluating how much additional education or training you're willing to pursue

Understanding Your Career Personality Type

Just as military service reveals certain aspects of your personality and working style, civilian career exploration benefits from understanding your natural preferences and strengths. Based on extensive research into personality types and career satisfaction, most people fall into one of five primary career personality categories. Understanding your type helps you target careers where you're likely to thrive and find satisfaction.

The Creative/Artist/Chef Personality

This personality type thrives on innovation, creation, and making a tangible impact on people's lives. Veterans with this personality often found satisfaction in military roles that involved training others, developing programs, or finding creative solutions to operational challenges.

Creative/Chef Characteristics:

- **Innovation-driven:** You enjoy finding new ways to solve problems or improve processes
- **People-focused:** You gain energy from helping, teaching, or serving others
- **Hands-on approach:** You prefer work where you can see tangible results from your efforts
- **Values-oriented:** You want your work to align with your personal values and beliefs
- **Artistic expression:** You may have interests in visual arts, music, writing, or design
- **Entrepreneurial spirit:** You might dream of starting your own business or nonprofit
- **Service continuation:** You want to continue serving others in civilian life

Military Experiences that Indicate Creative/Chef Personality:

- Enjoyed training and mentoring junior personnel
- Found satisfaction in community outreach or volunteer activities during service
- Participated in morale, welfare, and recreation (MWR) activities
- Developed creative solutions to operational challenges
- Enjoyed roles that involved direct interaction with people
- Felt energized by helping others solve problems or achieve goals
- Participated in religious or chapel services, community service projects

Civilian Career Paths for Creative/Chef Veterans:

Education and Training:

- Teacher (K-12, vocational, or adult education)
- Corporate trainer or instructional designer
- College professor or academic administrator
- Military instructor or contractor training positions

Healthcare and Social Services:

- Nurse, physician assistant, or healthcare administrator
- Social worker or counselor specializing in veteran issues
- Physical therapist or occupational therapist
- Mental health counselor or substance abuse counselor

Public Service and Nonprofit:

- Community activist or nonprofit director
- Social worker or case manager
- Public administrator or government relations
- Veteran service organization leadership

Entrepreneurship and Business:

- Restaurant or catering business owner
- Consulting firm focused on your military expertise
- Nonprofit organization founder
- Franchise owner in the service industry

Creative and Artistic Fields:

- Photographer specializing in military or family portraits
- Writer or journalist covering military/veteran topics
- Graphic designer or marketing professional
- Interior designer or landscape architect

The Corporate/Field General Personality

This personality type excels at leadership, strategic thinking, and managing complex operations. Veterans with this personality often held leadership positions in the military and found satisfaction in coordinating missions, managing personnel, and achieving organizational objectives.

Corporate/Field General Characteristics:

- **Leadership-oriented:** You naturally take charge and guide others toward objectives
- **Strategic thinking:** You enjoy planning, organizing, and seeing the big picture
- **Results-focused:** You measure success by concrete achievements and outcomes
- **Authority-comfortable:** You're comfortable making decisions and taking responsibility
- **Systems-minded:** You understand how different parts of organizations work together
- **Goal-driven:** You thrive when working toward specific, measurable objectives
- **Competition-motivated:** You enjoy challenges and competing to achieve excellence

Military Experiences that Indicate Corporate/Field General Personality:

- Held leadership positions (NCO, officer, supervisory roles)
- Enjoyed planning and executing complex operations
- Found satisfaction in managing budgets, resources, or personnel
- Thrived in competitive environments or military competitions
- Enjoyed roles that required coordination with multiple units or agencies
- Felt energized by achieving mission objectives and measurable results
- Participated in strategic planning or policy development

Civilian Career Paths for Corporate/Field General Veterans:

Corporate Leadership:

- Project manager or program manager
- Operations manager or director
- Sales manager or business development director
- Executive leadership roles (CEO, COO, VP positions)

Financial Services:

- Investment advisor or financial planner
- Insurance broker or agency owner
- Banking or credit union management
- Real estate investment or development

Government and Politics:

- Government administrator or department head
- Political campaign manager or elected official
- Law enforcement leadership (police chief, sheriff)
- Emergency management director

Consulting and Professional Services:

- Management consultant
- Business strategy advisor
- Government contractor or defense industry
- Legal services (attorney specializing in military/business law)

The Engineer/Technical Specialist Personality

This personality type excels at analysis, problem-solving, and working with systems or technology. Veterans with this personality often found satisfaction in military roles that involved technical operations, equipment maintenance, or analytical work.

Engineer/Technical Specialist Characteristics:

- **Problem-solving focused:** You enjoy diagnosing problems and finding technical solutions
- **Detail-oriented:** You pay attention to specifics and ensure accuracy in your work
- **Systems thinking:** You understand how technical systems work and interact
- **Independent work preference:** You often prefer working alone or in small technical teams
- **Continuous learning:** You enjoy mastering new technologies or technical skills
- **Quality-focused:** You take pride in doing technical work correctly and efficiently

- **Innovation through technology:** You see technology as a tool for improving processes

Military Experiences that Indicate Engineer/Technical Specialist Personality:

- Worked in technical MOS (information technology, electronics, mechanics, aviation)
- Enjoyed troubleshooting equipment or system problems
- Found satisfaction in maintaining or repairing complex equipment
- Preferred analytical or research-oriented assignments
- Enjoyed working with data, computers, or technical documentation
- Felt energized by mastering new technical skills or equipment
- Participated in technical training or certification programs

Civilian Career Paths for Engineer/Technical Specialist Veterans:

Engineering and Technology:

- Software developer or systems analyst
- Network administrator or cybersecurity specialist
- Mechanical, electrical, or civil engineer
- Quality assurance or process improvement specialist

Healthcare Technology:

- Biomedical technician or medical equipment specialist
- Health information technology specialist
- Radiologic technologist or medical imaging specialist
- Clinical research coordinator

Manufacturing and Industrial:

- Manufacturing engineer or process improvement specialist
- Maintenance supervisor or reliability engineer
- Quality control manager or inspection specialist
- Environmental health and safety specialist

The Accountant Personality

This personality type excels at precision, accuracy, and systematic organization of information. Veterans with this personality often found satisfaction in military roles that required meticulous attention to detail, financial management, or maintaining accurate records and documentation.

Accountant Personality Characteristics:

- **Detail-oriented:** You notice small discrepancies and ensure everything is accurate and complete
- **Systematic approach:** You prefer organized, step-by-step processes and clear procedures
- **Number-focused:** You enjoy working with financial data, statistics, and quantitative analysis
- **Compliance-minded:** You understand the importance of following regulations and standards
- **Quality assurance:** You take pride in producing error-free, accurate work
- **Risk-aware:** You naturally identify potential problems and work to prevent them
- **Process improvement:** You see opportunities to make systems more efficient and accurate
- **Quietly Accomplishes the Mission:** Your leadership is surprised when you complete assignments quickly without notice.

Military Experiences that Indicate an Accountant Personality:

- Worked in finance, administration, or logistics roles requiring precise documentation
- Enjoyed managing budgets, tracking expenses, and financial accountability
- Found satisfaction in maintaining accurate records and reports
- Preferred roles that required following detailed procedures and regulations
- Enjoyed quality control, inspection, or audit responsibilities
- Felt energized by organizing systems and ensuring compliance

- Participated in inventory management or supply chain operations

Civilian Career Paths for Accountant Personality Veterans:

Financial and Accounting Services:

- Certified Public Accountant (CPA) or bookkeeper
- Tax preparer or tax strategist
- Financial analyst or budget analyst
- Auditor (internal, external, or government)

Business Operations:

- Operations analyst or process improvement specialist
- Compliance officer or risk management specialist
- Project coordinator or administrative manager
- Quality assurance manager

Government and Regulatory:

- Government accountant or budget analyst
- Regulatory compliance specialist
- Contract administrator or procurement specialist
- Grant administrator or program analyst

Healthcare and Insurance:

- Medical coding specialist or health information manager
- Insurance claims adjuster or underwriter
- Healthcare financial analyst
- Benefits administrator

The Attorney/Historian Personality

This personality type excels at analyzing complex information, identifying patterns, and synthesizing large amounts of data into coherent conclusions. Veterans with this personality often found satisfaction in military roles that involved research, analysis, planning, or making sense of complicated situations.

Attorney/Historian Personality Characteristics:

- **Research-oriented:** You enjoy gathering information from multiple sources and analyzing it thoroughly
- **Pattern recognition:** You can identify trends, connections, and relationships in complex data
- **Critical thinking:** You naturally question assumptions and evaluate evidence
- **Communication skills:** You can explain complex information clearly to different audiences
- **Analytical mindset:** You break down complicated problems into manageable components
- **Detail synthesis:** You can see both the big picture and important details
- **Documentation focus:** You understand the importance of maintaining accurate records and precedents

Military Experiences that Indicate Attorney/Historian Personality:

- Worked in intelligence, planning, or strategic analysis roles
- Enjoyed researching regulations, policies, or procedures
- Found satisfaction in briefing leadership or preparing detailed reports
- Preferred roles that required analyzing multiple sources of information
- Enjoyed investigating incidents or conducting formal inquiries
- Felt energized by solving complex problems through research and analysis
- Participated in legal proceedings, boards, or administrative investigations

Civilian Career Paths for Attorney/Historian Personality Veterans:

Legal Services:

- Attorney specializing in military, veteran, or government law
- Paralegal or legal research specialist

- Court administrator or legal analyst
- Corporate counsel or compliance attorney

Research and Analysis:

- Policy analyst or research specialist
- Market research analyst or business intelligence specialist
- Data analyst or statistician
- Academic researcher or think tank analyst

Government and Public Service:

- Legislative analyst or policy advisor
- Government investigator or inspector
- Intelligence analyst (civilian agencies)
- Program evaluator or performance analyst

Business and Consulting:

- Management consultant or strategy advisor
- Risk analyst or due diligence specialist
- Corporate historian or knowledge management specialist
- Grant writer or proposal specialist

The Military Experience Integration Story

Understanding how your military experience integrates with civilian career choice is illustrated by the story of five veteran colleagues who started a comprehensive business services firm together, each contributing their unique personality type and military experience to create a successful enterprise.

Marcus - The Creative/Chef (Army Culinary Specialist turned Training Developer). Marcus served as a culinary specialist in the Army but discovered his passion for developing training programs and mentoring soldiers. After military service, he used his GI Bill to earn a degree in instructional design. He found his calling in creating engaging training programs that help people develop new skills. His military experience taught him the importance of clear, practical instruction that produces measurable results.

David - The Corporate/Field General (Marine Corps Officer turned Operations Director). David served as a Marine Corps officer, where he developed strong leadership, strategic planning, and business management skills. After earning an MBA, he built a successful career in corporate management. David's military experience in mission planning, resource management, and personnel leadership translated directly to business operations management and strategic planning.

Jason - The Engineer/Technical Specialist (Air Force Electronics Technician turned IT Director). Jason served as an electronics technician in the Air Force, where he developed expertise in troubleshooting complex systems and maintaining technical equipment. After military service, he transitioned into information technology and cybersecurity. His systematic approach to problem-solving and attention to technical detail made him invaluable in managing the firm's technology infrastructure.

Sarah - The Accountant (Navy Disbursing Clerk turned CFO). Sarah served as a disbursing clerk in the Navy, where she managed financial transactions, maintained accurate records, and ensured compliance with financial regulations. Her attention to detail and systematic approach to financial management made her a natural choice for managing the firm's finances, budgeting, and compliance requirements. She earned her CPA certification and became the firm's Chief Financial Officer.

Michael - The Attorney/Historian (Army Intelligence Analyst turned Legal Counsel). Michael served as an intelligence analyst in the Army, where he developed skills in researching complex information, analyzing patterns, and preparing detailed briefings for leadership. After using his GI Bill to attend law school, he specialized in business law and government contracts. His ability to analyze complex regulations and synthesize information from multiple sources made him essential for legal compliance and contract management.

The Power of Complementary Skills: The business services firm succeeded because each veteran contributed their unique strengths:

- Marcus provided training development and client education services
- David provided strategic planning and operational management
- Jason provided technology infrastructure and cybersecurity services
- Sarah provided financial management and compliance oversight
- Michael provided legal services and contract management

This story illustrates several important career principles for veterans:

1. **Know your strengths:** Each veteran understood what they did best and focused on those areas

2. **Recognize limitations:** Each veteran knew where they needed support and collaboration
3. **Leverage military experience:** All five applied lessons learned in military service to civilian business success
4. **Value teamwork:** Military experience taught them how to work together effectively toward common objectives
5. **Adapt skills:** They translated military skills to civilian contexts while maintaining military values

Finding Your Dream Career Match

Discovering your dream career requires honest self-assessment, exploration of options, and strategic decision-making that considers both your personality type and practical requirements.

Career Discovery Process:

Step 1: Self-Assessment

- Complete personality type assessment based on military experiences and preferences
- Identify your strongest skills and most enjoyable military experiences
- Consider what energizes you versus what drains your energy
- Evaluate your values and what gives your work meaning

Step 2: Market Research

- Research career options that match your personality type and interests
- Investigate salary ranges, job outlook, and geographic availability
- Identify education or certification requirements for target careers
- Connect with veterans working in fields that interest you

Step 3: Skill Gap Analysis

- Compare your current skills with the requirements for target careers
- Identify additional education, training, or certifications needed
- Evaluate whether you're willing to invest time and money in additional preparation
- Consider how to leverage military experience while filling skill gaps

Step 4: Practical Considerations

- Assess financial requirements and family obligations
- Consider geographic flexibility and family preferences
- Evaluate timeline for career transition and training
- Plan how to maintain income during any transition period

Step 5: Strategic Planning

- Develop a step-by-step plan for reaching your dream career
- Identify short-term positions that move you toward long-term goals
- Create a timeline with specific milestones and deadlines
- Build a network of mentors and supporters in your target field

Incorporating the Emotional Bucket Concept

Managing your Emotional Bucket is crucial for effective career exploration and decision-making. When your bucket is full of transition stress and uncertainty, it becomes difficult to make clear-headed career decisions or present yourself effectively to potential employers.

Impact of a Full Emotional Bucket on Career Development:

- Decision paralysis when overwhelmed by choices and uncertainty
- Poor interview performance when stress affects confidence and communication
- Unrealistic expectations that lead to disappointment and further stress
- Impatience with the process that leads to poor career decisions

Benefits of Managing Your Emotional Bucket:

- Clear thinking for practical career assessment and planning
- Confident presentation in networking and interview situations
- Realistic planning that accounts for challenges and timelines
- Sustained motivation for long-term career development efforts

Exercise

VETERAN CAREER DISCOVERY WORKSHOP

Objective: Identify your career personality type and develop a strategic plan for achieving your dream civilian career.

Part 1: Career Personality Assessment Rate how much you enjoyed these military experiences (1=Disliked, 5=Loved):

Creative/Chef Indicators:

- Training or mentoring others: ___
- Community service or volunteer work: ___
- Finding creative solutions to problems: ___
- Working directly with people: ___
- Participating in morale events: ___
- **Total Creative/Chef Score: ___/25**

Corporate/Field General Indicators:

- Leadership positions or responsibilities: ___
- Planning and executing operations: ___
- Managing budgets or resources: ___
- Competing in military competitions: ___
- Coordinating with multiple units: ___
- **Total Corporate/Field General Score: ___/25**

Engineer/Technical Specialist Indicators:

- Working with technical equipment: ___
- Troubleshooting system problems: ___
- Analyzing data or technical documentation: ___

- Working independently on projects: ___
- Learning new technical skills: ___
- **Total Engineer/Technical Specialist Score: ___/25**

Accountant Indicators:

- Managing financial records or budgets: ___
- Ensuring accuracy in documentation: ___
- Following detailed procedures and regulations: ___
- Quality control or inspection duties: ___
- Organizing systems and maintaining compliance: ___
- **Total Accountant Score: ___/25**

Attorney/Historian Indicators:

- Researching policies or regulations: ___
- Analyzing complex information from multiple sources: ___
- Preparing detailed reports or briefings: ___
- Investigating incidents or conducting inquiries: ___
- Identifying patterns or trends in data: ___
- **Total Attorney/Historian Score: ___/25**

Your Primary Personality Type: ______________________ (Highest score indicates primary type)

Part 2: Dream Career Exploration

If you were guaranteed success in any civilian career field, which career field would you choose if someone paid you $275,000+ per year to do it?

____

____

Why does this career appeal to you?

____

____

How does this career align with your personality type?

____

____

Part 3: Skills and Experience Inventory

Military Skills That Transfer to Your Dream Career:

1. ____________________________________
2. ____________________________________
3. ____________________________________
4. ____________________________________
5. ____________________________________

Additional Skills or Education Needed:

1. ____________________________________
2. ____________________________________
3. ____________________________________
4. ____________________________________
5. ____________________________________

Part 4: Career Development Plan

Short-term Goals (6-12 months):

- Research and networking goal: ____________________
- Education or training goal: ____________________
- Experience-building goal: ____________________

Medium-term Goals (1-2 years):

- Skill development goal: ____________________
- Position or role goal: ____________________
- Income or advancement goal: ____________________

Long-term Goals (3-5 years):

- Career achievement goal: ____________________
- Leadership or expertise goal: ____________________
- Impact or legacy goal: ____________________

Part 5: Action Steps and Timeline

Next 30 Days:

1. ____________________
2. ____________________
3. ____________________

Next 6 Months:

1. ____________________
2. ____________________
3. ____________________

Next Year:

1. ____________________
2. ____________________
3. ____________________

Self-Reflective Questions

1. Which aspects of my military experience brought me the most satisfaction, and how can I find similar satisfaction in civilian careers?

2. What prevents me from pursuing my dream career, and which of these obstacles are real versus perceived?

3. How can I leverage my military experience and personality type to create a competitive advantage in my target career field?

4. What am I willing to sacrifice or invest (time, money, effort) to achieve my dream career goals?

5. How does the state of my Emotional Bucket affect my ability to make clear career decisions and present myself effectively to potential employers?

Affirmations

1. "My military experience has prepared me with valuable skills and qualities that make me competitive in civilian careers."
2. "I have the ability to discover and achieve a civilian career that provides both financial security and personal fulfillment."
3. "I am willing to invest the time and effort necessary to build the civilian career I deserve."

Actionable Steps

1. **Complete Career Personality Assessment:** Use the assessment tool to identify your primary career personality type and understand how it relates to your military experience.
2. **Research Target Careers:** Investigate 3-5 specific careers that match your personality type and interests, including requirements, salary ranges, and job outlook.
3. **Develop Skill Enhancement Plan:** Identify the top 3 skills or credentials you need to develop for your target career and create specific plans for acquiring them.
4. **Build Professional Network:** Connect with at least three professionals working in your target field and seek informational interviews or mentoring relationships.
5. **Create Career Timeline:** Develop a realistic timeline with specific milestones for achieving your dream career, including interim positions that move you toward your long-term goal.

Building Your Dream Career Legacy

Discovering and pursuing your dream career is about more than just finding a job, it's about creating a civilian life that honors your military service while building the future you deserve. Your military experience provides a strong foundation of skills, values, and work ethic that gives you significant advantages in civilian career development.

The key is understanding how your personality type aligns with career opportunities, translating your military experience into civilian value, and developing a strategic plan that moves you systematically toward your goals. Remember that career development is often a marathon, not a sprint, and the patience and persistence you learned in military service will serve you well in civilian career building.

Your dream career should provide not only financial security but also a sense of purpose and mission that honors the service mentality you developed in the military. Whether you become a creative entrepreneur, a corporate leader, a technical specialist, a detail-oriented financial professional, or an analytical problem-solver, the goal is to find work that challenges you, rewards your efforts, and allows you to continue making a positive impact on the world.

By following a systematic approach to career discovery and development, you can build a civilian career that provides satisfaction, security, and a sense of purpose that you deserve after your military service. Your job is not just about what you do for work; it's about how you continue to serve others and make a difference in civilian life.

Notes

Entrepreneurship - Building Your Own Mission

Chapter

18

"The entrepreneur always searches for change, responds to it, and exploits it as an opportunity."

- Peter Drucker

Military service instilled in you the ability to execute missions under pressure, lead teams through uncertainty, and accomplish objectives with limited resources. These same skills form the foundation of successful entrepreneurship. However, transitioning from military service to business ownership requires careful planning, proper mentorship, and a clear understanding of both the opportunities and risks involved.

Your Emotional Bucket plays a critical role in entrepreneurial success. The stress of starting and running a business can quickly drain your emotional reserves if not properly managed. Successful veteran entrepreneurs learn to maintain emotional balance while navigating the challenges of business ownership, customer acquisition, financial management, and team leadership.

The Four Pillars of Business Success

Every successful business, regardless of industry or size, must master four fundamental pillars. Think of these as the critical mission components that must all function effectively for your business to thrive. Military experience has prepared you to understand systems thinking and the importance of each component working in harmony with the others.

Understanding the Four Pillars Framework

Just as military operations require coordination between intelligence, logistics, personnel, and operations, successful businesses require mastery of these four pillars:

1. **Product/Service Development** - What you offer and how you deliver value
2. **Business Operations** - The infrastructure and systems that support your business
3. **Marketing** - How you attract and acquire new customers
4. **Sales** - How you convert prospects into paying customers

Weakness in any pillar will limit your entire business, regardless of how strong the other pillars might be. As your business grows, each pillar requires increasing attention and often dedicated personnel to manage effectively.

Pillar 1: Product/Service Development

Your Value Creation Engine

This pillar encompasses everything related to what you're offering to the marketplace. For veterans, this often starts with translating your military expertise into civilian-relevant solutions or selling your military expertise back to the military as a government contractor.

Key Components:

- **Service Definition:** Clearly defining what you do and the outcomes you deliver
- **Quality Standards:** Establishing consistent delivery standards that exceed expectations
- **Continuous Improvement:** Regularly updating and improving your offerings based on client feedback
- **Intellectual Property:** Developing proprietary methods, processes, or tools that differentiate you
- **Innovation Pipeline:** Continuously developing new services or improving existing ones

Veteran Advantage in Product/Service Development: Your military experience provides significant advantages in this pillar:

- **Systems Thinking:** You understand how different components work together
- **Quality Focus:** Military training emphasizes precision and excellence
- **Problem-Solving:** You're trained to find solutions under pressure
- **Mission Focus:** You understand the importance of delivering results

Development Considerations:

- Can you clearly explain the transformation or outcome your service provides?
- Do you have documented processes that ensure consistent quality?
- How will you scale your service delivery as demand grows?
- What makes your approach unique or superior to alternatives?

Pillar 2: Business Operations

Your Infrastructure Foundation

Operations encompass all the behind-the-scenes systems and processes that make your business function smoothly. This is often where veteran-owned businesses excel due to military emphasis on logistics and systems.

Critical Operational Components:

Legal and Compliance Infrastructure:

- Business structure (LLC, S-Corp, C-Corp, etc.)
- EIN (Employer Identification Number)
- Business licenses and permits
- Insurance coverage
- Contract templates and legal documents

Financial Systems:

- Business banking accounts
- Business credit cards and credit establishment
- Bookkeeping and accounting systems
- Tax planning and compliance
- Financial reporting and analysis

Human Resources:

- Hiring and onboarding processes
- Employee handbooks and policies
- Payroll systems
- Performance management
- Training and development programs

Customer Service Operations:

- Client onboarding processes
- Communication protocols
- Issue resolution procedures
- Quality assurance systems
- Client retention strategies

Logistics and Fulfillment:

- Service delivery systems
- Project management processes
- Inventory management (if applicable)
- Technology infrastructure
- Backup and contingency plans

Operational Scaling Considerations: As your business grows, operational demands increase exponentially:

- **Solo Phase:** You handle most operations personally
- **Small Team Phase:** Basic systems and one administrative support person
- **Growth Phase:** Dedicated operations manager and specialized roles
- **Scale Phase:** Department heads and sophisticated systems

Pillar 3: Marketing

Your Customer Acquisition Engine

Marketing is how potential customers discover you exist and understand the value you provide. Many veterans underestimate the importance of consistent marketing, thinking that good work will speak for itself.

Essential Marketing Components:

Brand Development:

- Clear value proposition and messaging
- Professional visual identity
- Consistent brand voice and personality
- Differentiation from competitors

Digital Presence:

- Professional website with clear calls-to-action
- Search engine optimization (SEO)
- Social media presence on relevant platforms (Top two are LinkedIn and Instagram)
- Email marketing systems

Content Strategy:

- Educational content that demonstrates expertise
- Case studies and success stories
- Regular blog posts or articles
- Video content and webinars

Lead Generation:

- Lead magnets and free valuable resources
- Networking and referral systems
- Speaking engagements and thought leadership
- Strategic partnerships

Customer Acquisition Capacity Planning: A critical question many entrepreneurs fail to ask: "How many new customers can we effectively handle each week or month?"

Capacity Considerations:

- **Service Delivery Capacity:** How many clients can you serve well simultaneously?
- **Onboarding Capacity:** How many new clients can you properly onboard each month?
- **Quality Maintenance:** At what point does growth compromise service quality?
- **Resource Planning:** Do you have the team and systems to support growth?

Marketing Metrics to Track:

- Cost per lead acquisition
- Conversion rates from leads to customers
- Customer lifetime value
- Return on marketing investment
- Pipeline health and velocity

Pillar 4: Sales

Your Revenue Generation Engine

Sales is the process of converting interested prospects into paying customers. This pillar determines whether your marketing efforts translate into actual revenue.

Sales System Components:

Sales Process Design:

- Lead qualification criteria
- Sales funnel stages and conversion points
- Follow-up sequences and timing
- Objection handling strategies
- Closing techniques and procedures

Pricing Strategy:

- Value-based pricing models
- Service packages and options
- Payment terms and policies
- Upselling and cross-selling opportunities

Sales Channel Management:

- Direct sales vs. online sales funnels
- Phone sales (Zoom calls) vs. in-person meetings
- Automated vs. high-touch sales processes
- Referral and partnership sales

Technology and Tools:

- Customer Relationship Management (CRM) systems
- Sales tracking and analytics
- Proposal and contract generation
- Payment processing systems

Sales Scaling Considerations:

Solo Sales Phase: You handle all sales personally, which works for:

- High-value, complex services requiring expertise
- Relationship-based businesses
- Custom solutions requiring consultation
- Services requiring significant trust-building

Automated Sales Phase: Online sales funnels that convert without human interaction, suitable for:

- Standardized products or services
- Lower-price-point offerings
- Educational products or courses
- Services with clear, predictable outcomes

Sales Team Phase: Dedicated salespeople when you need:

- Higher sales volume than you can personally handle
- Specialized sales skills for complex products
- Coverage across different territories or markets
- Focus on other business activities

Price Point and Sales Process Correlation:

Lower Price Points ($99-$499):

- Usually require automated online sales processes
- Minimal human interaction needed
- Clear value proposition and simple decision
- Streamlined purchase process

Medium Price Points ($500-$2,999):

- May work with automated systems or brief consultation
- Often benefit from FAQ pages and detailed information
- Some customers prefer speaking to someone before purchasing
- Email sequences and retargeting can be effective

Higher Price Points ($3,000+):

- Almost always require human interaction
- Multiple touchpoints and relationship building
- Detailed consultation and customization
- Trust and credibility are paramount

Sales Capacity and Growth Planning: As your business scales, consider:

- Can you sell a $999 coaching program without customer interaction?
- Do you need dedicated salespeople for higher-value services?
- How many sales conversations can you personally handle weekly?
- What's your conversion rate from consultation to sale?

The Veteran Entrepreneurship Advantage

Military Skills That Transfer to Business Building:

Mission-Focused Mindset: Your military training taught you to identify objectives, develop strategies, and execute plans to completion. This mission-focused approach translates directly to business goal-setting and achievement across all four pillars.

Leadership Under Pressure: Military leaders regularly make critical decisions with incomplete information and tight deadlines. Entrepreneurship requires similar decision-making capabilities when facing market changes, customer demands, and competitive pressures across all business areas.

Systems Thinking: Military operations require understanding how different components work together. This translates perfectly to managing the four pillars of business and understanding their interdependencies.

Resourcefulness and Adaptability: Military operations often require improvisation and creative problem-solving with limited resources. These skills are essential for entrepreneurs, especially in the early stages of building each business pillar.

Team Building and Development: Your experience building cohesive, high-performing teams translates to recruiting, training, and managing employees across all business functions.

Building Your Four-Pillar Business Strategy

Phase 1: Foundation Building (Months 1-6)

Product/Service Development:

- Define your core ***service*** or ***product*** offering and unique value proposition
- Develop standard operating procedures for service delivery
- Create quality control processes and metrics
- Document your methodology and approach

Business Operations:

- Establish legal business structure and obtain EIN
- Set up business banking and credit accounts
- Implement basic bookkeeping and financial systems
- Obtain necessary licenses and insurance

Marketing:

- Build a professional website with a clear value proposition
- Establish social media presence on 1-2 key platforms
- Create lead magnet and email capture system (Sales funnel)
- Develop content strategy and begin regular publishing

Sales:

- Design sales process and qualification criteria
- Create pricing strategy and service packages
- Develop proposal templates and contracts
- Establish payment processing systems

Phase 2: Growth and Systematization (Months 6-18)

Product/Service Development:

- Refine service offerings based on client feedback
- Develop additional services or packages (Once you have generated substantial income with the first offering)
- Create training materials and documentation
- Begin developing proprietary tools or methods

Business Operations:

- Hire first team member (typically administrative support)
- Implement customer service protocols
- Establish performance measurement systems
- Create an employee handbook and policies

Marketing:

- Expand content marketing and thought leadership
    - Consider your own YouTube Channel to discuss your view in the industry
- Develop case studies and testimonials
- Build referral and partnership networks
- Increase your marketing budget and sophistication

    - Consider running paid ads on social media platforms to connect with B2C prospects
    - Consider attending your industry association conferences to meet B2B prospects
    - For B2G, it's essential that you meet the government contracting person
        - We sell to people and not to buildings – personality/connection matters

Sales:

- Optimize sales process and improve conversion rates
- Consider automation for lower-value offerings
- Develop sales scripts and objection handling
- Track and analyze sales metrics

Phase 3: Scaling and Optimization (Months 18+)

Product/Service Development:

- Launch additional service lines or products
- Develop licensing or certification programs
- Create passive income streams
- Build innovation pipeline

Business Operations:

- Hire specialized team members for each pillar
- Implement advanced systems and technologies
- Establish multiple office locations (if applicable)
- Develop succession and exit planning

Marketing:

- Build a comprehensive marketing funnel
- Expand to additional marketing channels
- Develop brand partnerships and collaborations
- Measure and optimize marketing ROI

Sales:

- Build a dedicated sales team
- Implement sophisticated CRM and analytics
- Develop multiple sales channels
- Create scalable sales processes

Common Veteran Entrepreneurship Mistakes

Pillar Neglect - Many veteran entrepreneurs excel in one or two pillars while neglecting others:

- **Operations-Heavy:** Great systems but poor marketing and sales
- **Product-Focused:** Excellent service but inadequate business infrastructure
- **Sales-Driven:** High revenue but poor operations and service quality

Jumping Too Quickly - The biggest mistake veteran entrepreneurs make is rushing into business ownership without properly building all four pillars. This leads to:

- Customer service problems due to poor operations
- Revenue inconsistency due to weak marketing and sales
- Quality issues due to inadequate service development
- Financial problems due to poor business infrastructure

Underestimating Civilian Business Dynamics - Military contracts and government procurement operate differently from civilian markets:

- Customer acquisition requires relationship building and marketing
- Sales cycles may be longer and more complex
- Competition operates differently from military procurement
- Customer service expectations differ significantly

Financial Planning for Four-Pillar Development

Investment Requirements by Pillar:

Product/Service Development (15-20% of startup budget):

- Research and development time
- Training and certification costs
- Tool and resource development
- Quality assurance systems

Business Operations (40-50% of startup budget):

- Legal setup and compliance costs
- Technology infrastructure
- Office setup and equipment
- Initial staffing costs

Marketing (20-30% of startup budget):

- Website development and design
- Content creation and advertising
- Marketing tools and software
- Brand development costs

Sales (10-15% of startup budget):

- CRM and sales tools
- Proposal and contract systems
- Sales training and development
- Commission and incentive structures

Exercise

FOUR-PILLAR BUSINESS ASSESSMENT

Objective: Evaluate your readiness across all four business pillars and develop a strategic plan for building a comprehensive, scalable business that leverages your military experience while protecting your family's financial security. ***Use ChatGPT to assist you.***

Part A: Pillar 1 - Product/Service Development Assessment

1. Clearly define your core service offering and the specific transformation you provide
2. List the unique advantages your military experience gives you in this area
3. Describe your quality control processes and how you ensure consistent delivery
4. Identify how you'll scale service delivery as demand grows
5. Outline your plans for continuous improvement and innovation

Part B: Pillar 2 - Business Operations Assessment

1. List all legal and compliance requirements you need to address
2. Evaluate your current financial systems and identify gaps
3. Assess your customer service processes and capabilities
4. Determine your operational capacity limits and scaling needs
5. Identify which operational functions you'll need to hire first

Part C: Pillar 3 - Marketing Assessment

1. Define your target customer and where they can be found
2. Evaluate your current marketing presence and identify weaknesses
3. Determine how many new customers you can effectively handle monthly
4. Assess your content strategy and thought leadership approach
5. Calculate your customer acquisition costs and capacity

Part D: Pillar 4 - Sales Assessment

1. Design your sales process from lead to closed sale
2. Determine appropriate pricing for your service level and market
3. Evaluate whether your services can be sold online without human interaction
4. Assess your personal sales capacity and when you'll need help
5. Calculate your conversion rates and sales cycle length

Part E: Integration and Scaling Strategy

1. Identify which pillar is currently your strongest and weakest
2. Create a timeline for building and strengthening each pillar
3. Determine when you'll need to hire specialists for each area
4. Plan how the four pillars will work together as you scale
5. Establish metrics for measuring success in each pillar

Self-Reflective Questions

1. How can I leverage my military experience to build advantages in each of the four business pillars while serving civilian customers effectively?

2. Which of the four pillars represents my greatest strength, and how can I use that strength to build credibility while developing the other pillars?

3. How will I ensure I'm building adequate capacity in marketing and sales to support the growth my operations and service delivery can handle?

4. What specific mentorship and guidance do I need for the business pillars where I have the least experience or confidence?

5. How will I maintain balance across all four pillars as my business grows, avoiding the common mistake of over-developing one pillar while neglecting others?

Affirmations

1. "I am building a comprehensive business that excels in all four pillars: exceptional service delivery, solid operations, effective marketing, and strong sales systems."
2. "I leverage my military training to create systematic approaches to business building that ensure sustainable growth and excellent customer service."
3. "I invest time and resources in developing each business pillar because I understand that my long-term success depends on balanced, comprehensive business development."

Actionable Steps

1. **Conduct Four-Pillar Assessment**: Within the next two weeks, complete a thorough assessment of your current status in each pillar and identify your three biggest gaps.
2. **Create Pillar Development Timeline**: Within 30 days, develop a 12-month plan for systematically building and improving each of the four pillars with specific milestones and deadlines.
3. **Secure Pillar-Specific Mentorship**: Within 60 days, identify and connect with mentors or advisors who can provide guidance for your weakest pillars, particularly in areas where military experience provides less advantage.
4. **Build Capacity Planning System**: Create a system for tracking your capacity in each pillar so you can scale appropriately and avoid overwhelming any single area of your business.
5. **Establish Pillar Metrics**: Set up measurement systems for each pillar so you can track progress and ensure balanced development across all areas of your business.

Conclusion

Entrepreneurship offers veterans an opportunity to continue serving others while building financial independence and professional fulfillment. Your military experience provides valuable advantages in the systematic approach required to build all four pillars of business success.

However, entrepreneurship also involves significant risks that require careful planning, adequate preparation, and ongoing support. The key to veteran entrepreneurial success lies in building comprehensive businesses that excel across all four pillars: exceptional product/service development, solid business operations, effective marketing, and strong sales systems.

Remember that building a successful business is a marathon, not a sprint. Take time to develop each pillar systematically, validate your market demand, secure adequate mentorship, and maintain the financial security your family depends on. Your military training taught you that proper preparation and execution lead to mission success; apply these same principles to your entrepreneurial mission.

Don't fall into the trap of developing one or two pillars while neglecting others. Businesses that scale successfully are those that maintain balance across all four areas, hiring specialists as you grow, and keeping the systematic approach that military service taught you.

Your entrepreneurial mission is an extension of your service: serving customers, creating value, and building something meaningful. Approach it with the same discipline, planning, and commitment to excellence that made you successful in the military.

Notes

Continuing Service - Community Impact, Board Membership, & Politics

Chapter

19

The transition from military service to civilian life often includes a continued desire to serve, not just personally, but at broader community, state, or national levels. Veterans bring unique qualifications to politics, public service, and community leadership: proven leadership under pressure, experience managing complex operations, a deep understanding of national security issues, and demonstrated commitment to service above self. This chapter explores how veterans can leverage their military experience in political roles, community board service, and public service while providing practical guidance for those considering these paths of continued service.

The Veteran's Call to Continued Service

Military service instills a deep commitment to serving something greater than yourself, a value that doesn't end with separation or retirement. Many veterans find that civilian life feels incomplete without a meaningful way to continue serving their communities. The transition often includes what researchers call "moral injury", not from traumatic events, but from the sudden absence of clear purpose and meaningful service that military life provided.

Why Veterans Seek Continued Service:

Mission-Oriented Identity: Military service creates an identity built around mission accomplishment and service to others, making purely personal pursuits feel insufficient.

Values Alignment: Military values of service, integrity, and commitment to others create a need for civilian activities that honor these same principles.

Leadership Experience: Veterans possess leadership skills that they want to use for positive impact rather than letting them atrophy in civilian life.

Community Connection: Military service creates bonds through shared sacrifice; veterans often seek civilian communities that provide a similar connection through shared service.

Skills Utilization: Veterans want to use their military-developed skills for positive impact rather than pursuing careers that don't leverage their full capabilities.

PART I: VETERANS IN POLITICS AND PUBLIC SERVICE

Why Veterans Excel in Politics and Public Service

Veterans possess distinctive qualities that translate exceptionally well to political and public service roles, making them natural candidates for elected office and public sector leadership.

Core Military-to-Political Advantages:

Leadership Under Pressure: Military service develops leaders who can make critical decisions under intense pressure, manage crises effectively, and maintain composure during heated debates or challenging public scrutiny.

Operational Planning: Veterans understand strategic planning, resource allocation, and logistics management, skills directly applicable to policy development, budget management, and government operations.

Diverse Coalition Building: Military experience working with people from different backgrounds, regions, and perspectives prepares veterans for the coalition-building essential in politics.

National Security Expertise: Veterans bring firsthand understanding of defense issues, international relations, and homeland security challenges that civilian politicians often lack.

Servant Leadership: The military's emphasis on serving something greater than yourself aligns perfectly with the public service mission of elected office.

Institutional Respect: Voters often view military service as evidence of character, commitment, and leadership capability, providing veterans with built-in credibility.

Types of Political and Public Service Opportunities

Veterans can pursue various paths in politics and public service, from local community involvement to national office.

Elected Positions:

- **Local Government:** City council, mayor, county commissioner, school board
- **State Government:** State legislature, governor, lieutenant governor, attorney general
- **Federal Government:** U.S. House of Representatives, U.S. Senate, President

Appointed Positions:

- **Federal Agencies:** Department of Defense, Veterans Affairs, Homeland Security, State Department
- **State Agencies:** Emergency management, veterans affairs, public safety
- **Local Government:** City manager, police chief, fire chief, department heads

Political Support Roles:

- **Campaign Management:** Managing campaigns for other candidates
- **Policy Advocacy:** Working for think tanks, advocacy organizations, or lobbying firms
- **Political Consulting:** Providing strategic advice to candidates and elected officials

Public Service Careers:

- **Civil Service:** Federal, state, and local government career positions
- **Law Enforcement:** Police, federal agents, investigators
- **Emergency Management:** Disaster response, homeland security, public safety

Veteran Organizations Supporting Political Campaigns

Several organizations specifically support veterans running for political office, providing training, funding, and strategic assistance that can make the difference between winning and losing campaigns.

National Organizations:

Veterans Campaign (VetsCampaign): A nonpartisan organization that provides campaign training, strategic guidance, and ongoing support to military veterans running for office at all levels. They offer:

- Campaign workshops and boot camps
- One-on-one mentoring from experienced political operatives
- Networking opportunities with other veteran candidates
- Resources for fundraising and voter outreach
- Post-election support for newly elected veterans

With Honor Action: A cross-partisan organization that supports principled veterans running for Congress. They provide:

- Direct financial support through their PAC
- Strategic campaign consulting
- Communications training and message development
- Coalition building and endorsement assistance
- Leadership development for veteran candidates

Iraq and Afghanistan Veterans of America (IAVA): While primarily a policy advocacy organization, IAVA also:

- Endorses veteran candidates who support their policy priorities
- Provides platform development assistance
- Offers media training and public speaking coaching
- Connects candidates with their extensive veteran network

Student Veterans of America (SVA): Supports veterans in various transitions, including politics:

- Leadership development programs
- Public speaking and advocacy training
- Networking opportunities with veteran professionals
- Educational resources on policy and governance

Party-Specific Organizations:

Republican Side:

- **Republican Veterans Coalition:** Provides campaign support, fundraising assistance, and strategic guidance for Republican veteran candidates

Democratic Side:

- **VoteVets:** Progressive organization supporting Democratic veteran candidates with:
    - Direct campaign contributions
    - Independent expenditure campaigns
    - Media production and strategic communications
    - Grassroots organizing support

Common Ground Coalition: A bipartisan group focused on supporting veterans who emphasize collaboration and problem-solving over partisan politics.

State and Local Veteran Political Organizations:

Many states have veteran political organizations that provide local support:

- **State Veterans Caucuses:** Bipartisan groups of veteran legislators who work together on veteran issues
- **Local Veteran Political Clubs:** Community-based organizations that support veteran candidates
- **Regional PACs:** Political action committees focused on supporting veterans in specific geographic areas

Getting Started in Politics and Public Service

For veterans considering political or public service careers, success requires strategic planning and gradual skill development.

Building Your Foundation:

Start Local: Begin with community involvement through:

- Volunteer work with local nonprofits
- Attendance at city council or school board meetings
- Participation in community organizations and civic groups
- Leadership roles in veteran service organizations

Develop Political Skills:

- Take public speaking and communication courses
- Attend political training programs offered by veteran organizations
- Volunteer for other candidates' campaigns to learn the process
- Study local, state, and national issues relevant to your interests

Build Your Network:

- Connect with other veteran politicians and public servants
- Join veteran political organizations
- Attend political events and fundraisers
- Develop relationships with party officials and community leaders

Establish Your Platform:

- Identify issues you're passionate about and knowledgeable about
- Develop policy positions based on your military experience and community needs
- Create a personal brand that highlights your service and vision
- Build a social media presence that demonstrates your engagement and expertise

PART II: COMMUNITY BOARD SERVICE AND NONPROFIT LEADERSHIP

Understanding Community Impact Opportunities

Veterans can create community impact through various channels, each offering different ways to serve while building civilian connections and purpose.

Types of Community Service Opportunities:

Nonprofit Board Service:

- **Local Nonprofits:** Community organizations addressing local needs like homelessness, education, inner-city youth development (technical skills, emotional wellness, relationships, education, sports, etc.), or economic development
- **Veteran Service Organizations:** Groups specifically serving veteran communities and families
- **National Organizations:** Larger nonprofits with local chapters addressing national issues
- **Issue-Specific Organizations:** Groups focused on causes you're passionate about, such as youth development, environmental protection, or health advocacy

Educational Board Service:

- **School Boards:** Local public school governance providing direct community impact
- **Charter School Boards:** Innovation-focused educational leadership opportunities
- **Community College Boards:** Workforce development and community education oversight
- **University Boards:** Higher education governance and strategic direction

Healthcare and Social Services:

- **Hospital Boards:** Healthcare system governance and community health improvement
- **Mental Health Organizations:** Addressing community mental health needs, particularly veteran-focused services
- **Social Service Agencies:** Organizations addressing poverty, housing, family services, and community welfare
- **Public Health Initiatives:** Community health improvement and emergency preparedness

Economic and Community Development:

- **Economic Development Boards:** Attracting business and improving local economic conditions
- **Chamber of Commerce:** Business community leadership and economic advocacy
- **Housing Authorities:** Addressing community housing needs and urban development
- **Regional Planning Organizations:** Long-term community planning and infrastructure development

Civic and Government Service:

- **City Planning Commissions:** Community development and land use planning
- **Parks and Recreation Boards:** Community recreation and quality of life improvement
- **Library Boards:** Community education and information access
- **Public Safety Commissions:** Community safety and emergency preparedness oversight

The Value Veterans Bring to Board Service

Veterans possess unique qualifications that make them exceptional board members and community leaders.

Military-Developed Board Skills:

Strategic Planning Experience: Military training in mission planning, resource allocation, and strategic thinking directly applies to nonprofit and community organization governance.

Financial Management: Military experience with budgets, resource optimization, and accountability systems translates well to board fiduciary responsibilities.

Crisis Management: Veterans' experience managing crises and making decisions under pressure provides a valuable perspective for organizational challenges.

Team Leadership: Military leadership experience helps boards function more effectively through improved communication, decision-making, and conflict resolution.

Mission Focus: Veterans understand the importance of staying focused on the organizational mission and measuring success against clear objectives.

Ethical Standards: Military training in integrity and ethical decision-making helps organizations maintain high standards and public trust.

Diverse Perspective: Veterans bring unique perspectives on service, sacrifice, and commitment that can enhance board discussions and decision-making.

Community Credibility: Military service often provides automatic credibility and trust that helps organizations build community support and partnerships.

Veteran-Specific Service Opportunities

Veterans Service Organizations:

- **American Legion, VFW, DAV:** Traditional veteran service organizations with local, state, and national board opportunities
- **Iraq and Afghanistan Veterans of America (IAVA):** Policy advocacy and veteran support organization
- **Team Rubicon:** Disaster response organization that leverages veteran skills
- **The Mission Continues:** Community service organization connecting veterans with volunteer opportunities

Military Family Support:

- **Blue Star Families:** Supporting military families through deployment and transition
- **Operation Homefront:** Providing emergency assistance and support to military families
- **Fisher House Foundation:** Supporting families during medical treatment at military hospitals
- **Armed Services YMCA:** Serving military families with programs and support services

Education and Youth Development:

- **Student Veterans of America:** Supporting veteran students in higher education
- **Veterans Education Success:** Advocating for veteran education benefits and support
- **Boys and Girls Clubs:** Youth development organizations that value military leadership
- **Junior ROTC Support Organizations:** Supporting youth military leadership development

Community Impact and Leadership Development:

- **Leadership Development Programs:** Community leadership initiatives that value military experience
- **Rotary, Kiwanis, Lions Club:** Service organizations focused on community improvement
- **United Way:** Community impact organizations addressing multiple social issues
- **Habitat for Humanity:** Housing-focused service organization

PART III: INTEGRATED APPROACH TO CONTINUED SERVICE

Building Your Service Portfolio

Successful veteran community leaders often combine multiple forms of service to maximize their impact and build comprehensive leadership experience.

Progressive Engagement Strategy:

Phase 1: Foundation Building (Months 1-12)

- Begin with volunteer service in areas of interest
- Attend community meetings and events
- Join veteran service organizations
- Take leadership training courses
- Build relationships with community leaders

Phase 2: Active Engagement (Years 1-3)

- Secure first board appointment
- Take on committee leadership roles
- Volunteer for political campaigns
- Speak at community events
- Mentor other veterans in service

Phase 3: Leadership Development (Years 3-5)

- Serve on multiple complementary boards
- Consider running for local office
- Take on board chair or officer roles
- Lead community initiatives
- Build a reputation as an effective leader

Phase 4: Expanded Impact (Years 5+)

- Run for higher office or major appointments
- Serve on regional or national boards
- Become a recognized community leader
- Mentor next generation of veteran leaders
- Create lasting institutional change

Campaign Strategy for Veterans

Whether running for office or leading community organizations, veterans can leverage common strategic advantages.

Messaging Advantages:

- Emphasize service, sacrifice, and commitment to community
- Highlight leadership experience and decision-making under pressure
- Connect military problem-solving skills to civilian challenges
- Demonstrate bipartisan appeal through military service

Avoiding Common Pitfalls:

- Don't assume military service alone qualifies you for office or board service
- Avoid military jargon that civilians may not understand
- Balance military experience with civilian concerns and priorities
- Address any controversial aspects of your military service proactively

Building Support Networks:

- Connect with veteran political and service organizations early
- Apply for campaign and leadership training programs
- Seek endorsements from veteran groups and community leaders
- Utilize veteran networks for volunteers, donations, and support

Board Service - Best Practices for Veterans

Successful board service requires adapting military leadership skills to civilian governance contexts.

Effective Board Participation:

1. Governance vs. Management:

- Understand that boards govern and set policy while staff manage day-to-day operations
- Avoid micromanaging or trying to run operations unless specifically asked
- Focus on strategic direction, oversight, and support rather than tactical execution

2. Collaborative Leadership:

- Adapt military command leadership to collaborative civilian board leadership
- Listen to other board members' perspectives and expertise before offering your own
- Build consensus and support for decisions rather than expecting orders to be followed

3. Financial Oversight:

- Use military budget experience to understand organizational finances
- Ask questions about financial management and sustainability
- Support fundraising efforts even if this wasn't part of your military experience

Incorporating the Emotional Bucket Concept

Political campaigns, public service careers, and board service can be emotionally demanding, making emotional bucket management crucial for veterans in these fields.

Impact of a Full Emotional Bucket on Service Performance:

- **Increased Reactivity:** Overwhelming emotions can lead to inappropriate responses during debates, board meetings, or public appearances.
- **Poor Decision-Making:** Stress from unprocessed military experiences can cloud judgment on policy decisions or organizational governance.
- **Communication Problems:** Emotional overload can affect the ability to connect with voters, fellow board members, or community stakeholders.
- **Burnout Risk:** Service stress combined with unmanaged military trauma can lead to rapid burnout and ineffective leadership.

Benefits of Managing Your Emotional Bucket:

- **Enhanced Public Communication:** Clear emotional state improves communication with constituents, board members, and media.
- **Better Leadership:** Emotional wellness enables more effective leadership and coalition building.
- **Improved Relationships:** Well-managed emotions strengthen relationships with staff, colleagues, voters, and community partners.
- **Sustained Performance:** Emotional health supports long-term effectiveness in demanding service roles.

Exercise

INTEGRATED SERVICE READINESS ASSESSMENT

Objective: Evaluate your readiness for various forms of continued service and create a comprehensive development plan that may include politics, board service, and community leadership.

1. Service Vision and Goals

Long-term Service Vision: Describe your ideal service contribution 5-10 years from now:

____

____

____

Service Interests Assessment: Rate your interest level (1-5) in each area:

- _____ Local elected office
- _____ State/federal elected office
- _____ Appointed government positions
- _____ Nonprofit board service
- _____ Educational board service
- _____ Veterans organization leadership
- _____ Community organization leadership
- _____ Policy advocacy and lobbying
- _____ Campaign management/consulting

Primary Service Goals:

1. ____
2. ____
3. ____
4. ____

2. Skills and Experience Assessment

Military Leadership Background:

- _____ Years of service
- _____ Highest rank
- _____ Leadership positions
- _____ Budget management
- _____ Crisis management examples

Civilian Leadership Development:

- _____ Board/committee experience
- _____ Volunteer leadership
- _____ Political involvement
- _____ Public speaking experience
- _____ Community connections

Transferable Skills Inventory: Rate your current level (1-5) in each area:

- _____ Strategic planning
- _____ Financial management
- _____ Public speaking
- _____ Fundraising
- _____ Coalition building
- _____ Crisis management
- _____ Media relations
- _____ Policy analysis

3. Political and Community Landscape Analysis

Your Community Assessment:

- Population size ______________________________
- Major political issues ______________________________
- Key community needs ______________________________
- Veteran population ______________________________
- Political climate ______________________________

Potential Opportunities:

- Open or upcoming elections ______________________________
- Board openings you're aware of ______________________________
- Community organizations needing leadership: ______________________________
- Issues lacking veteran voice ______________________________

Political Positioning:

- Party affiliation (if any) ______________________________
- Key policy positions ______________________________
- Potential support base ______________________________
- Likely challenges ______________________________

4. Support Network and Resources

Veteran Organization Connections:

- ☐ Connected with Veterans Campaign
- ☐ Engaged with "With Honor Action"
- ☐ Member of IAVA or SVA
- ☐ Involved with state veteran political organizations
- ☐ Connected with local veteran groups
- ☐ Active with VFW, American Legion, or DAV

Community Networks:

- ☐ Attended political events
- ☐ Volunteered for campaigns
- ☐ Connected with party officials
- ☐ Built relationships with community leaders
- ☐ Established nonprofit connections
- ☐ Developed donor relationships

Professional Support Needed:

- ☐ Campaign training
- ☐ Board governance education
- ☐ Public speaking coaching
- ☐ Fundraising training
- ☐ Media training

5. Development Plan Creation

Immediate Steps (Next 6 Months):

Skill Development Priorities:

1. ____________________
2. ____________________
3. ____________________
4. ____________________

Organization Connections to Make:

1. ____________________
2. ____________________
3. ____________________
4. ____________________

Community Involvement to Pursue:

1. ______________________________

2. ______________________________

3. ______________________________

Training Programs to Attend:

1. ______________________________

2. ______________________________

Medium-term Goals (6 months to 2 years):

Leadership Positions to Seek:

1. ______________________________

2. ______________________________

3. ______________________________

Campaign or Board Experience to Gain:

1. ______________________________

2. ______________________________

3. ______________________________

Policy Expertise to Develop:

1. ______________________________

2. ______________________________

3. ______________________________

Network Expansion Targets:

1. ______________________________

2. ______________________________

3. ______________________________

Long-term Objectives (2+ years):

Target Offices or Positions:

1. ______________________________
2. ______________________________
3. ______________________________

Timeline for Major Goals:

- First board appointment ____________________
- First campaign involvement ____________________
- Potential candidate timeline ____________________
- Major leadership role timeline ____________________

6. Resource and Support Requirements

Financial Considerations:

- Personal financial capacity for service ______________
- Fundraising needs for potential campaigns ______________
- Board contribution expectations ______________
- Professional development budget ______________

Time Management:

- Available hours per week for service ____________________
- Family considerations ____________________
- Career impact assessment ____________________
- Long-term commitment capacity ____________________

Family and Personal Support:

- ☐ Acquired family support for service goals
- ☐ Considered impact on personal relationships
- ☐ Developed a stress management plan
- ☐ Developed a strategy for work-life balance

Self-Reflective Questions

1. How can I best use my military experience to serve my community through politics, board service, or other leadership roles?

2. What form of continued service would provide the greatest sense of purpose and mission fulfillment?

3. How does the state of my Emotional Bucket affect my readiness for the demands of public service or community leadership?

4. What barriers might prevent me from pursuing my service goals, and how can I address them?

5. How can I build the skills and relationships necessary for effective service while maintaining my other responsibilities?

Affirmations

1. My military service has prepared me to serve my community and country in new and meaningful ways.

2. I have valuable leadership experience and skills that can make a positive difference in my community.

3. I am committed to continued service that honors my military values while meeting civilian community needs.

Actionable Steps

1. **Choose Your Service Path:** Based on your assessment, identify whether you're most interested in politics, board service, or community leadership, and focus initial efforts accordingly.

2. **Connect with Veteran Service Organizations:** Reach out to organizations that support veterans in your chosen service area and begin building relationships.

3. **Start Community Engagement:** Begin attending local government meetings, nonprofit events, or political gatherings to understand your community's needs and dynamics.

4. **Develop Required Skills:** Enroll in training programs for campaign management, board governance, or public speaking based on your service goals.

5. **Build Your Network:** Systematically build relationships with current leaders in your areas of interest, including both veterans and civilians who can mentor and support your service goals.

The Integrated Path to Continued Service

Veterans have unique opportunities to continue serving their communities through politics, board service, and community leadership. Your military experience provides valuable preparation for these roles, but success requires adapting your skills to civilian contexts and building new capabilities specific to your chosen form of service.

Whether your path leads to elected office, nonprofit board leadership, or community organization management, the key is to start with volunteer service, build civilian relationships and skills, and gradually take on increasing responsibility. The veteran support organizations and training programs available can significantly accelerate your progress and increase your chances of making meaningful impact.

Remember that effective service is a long-term commitment that builds over time. Focus on making genuine contributions to your community while building the relationships and reputation necessary for broader leadership opportunities. Your military service instilled values of commitment, integrity, and service that will serve you well in any form of continued service you choose to pursue.

Evidence-Based Research: Studies show that veterans in elected office often demonstrate higher levels of bipartisan cooperation and focus on problem-solving rather than partisan politics (Bipartisan Policy Center, 2018). Research also indicates that veterans who engage in meaningful community service report higher levels of life satisfaction and lower rates of depression during civilian transition (Matthieu et al., 2012). Additionally, veterans serving on nonprofit boards bring unique value through their leadership experience, strategic thinking, and commitment to mission accomplishment (BoardSource, 2019).

Notes

Cultivating Resilience – From Military to Civilian Challenges

Chapter

20

Resilience, the ability to bounce back from adversity and continue moving forward despite setbacks, is deeply embedded in military culture. However, the resilience that served you in uniform must be fundamentally adapted for civilian life, where the comprehensive institutional support system that once handled every aspect of your existence has vanished. This chapter explores how to rebuild resilience when the safety net is gone and you must create your own support systems from scratch.

The Reality of Lost Support Systems

Military service provided an all-encompassing life support ecosystem that most veterans don't fully appreciate until it's gone. The military handled your food through mess halls, your clothing through uniforms and allowances, your shelter through barracks and base housing, and your transportation through military vehicles and travel orders. Your healthcare was automatic, your fitness was structured and supervised, and your mental health support was integrated into the system. Even your spiritual needs were addressed through chaplain services, while the leadership structure provided clear guidance and decision-making frameworks.

The military's support extended far beyond basic needs to include guaranteed paychecks with predictable raises, world travel in controlled and safe environments, built-in social networks of comrades who shared your experiences, family support programs, shopping and travel discounts, and comprehensive safety and security measures. Your career progression was mapped out, your training was provided, and your social connections were automatically built through shared missions and unit cohesion. This institutional cocoon allowed you to focus on mission completion while everything else in life was managed for you.

When you leave military service, this comprehensive support system disappears overnight. Suddenly, you're responsible for securing housing, finding healthcare providers, building social networks, managing finances without guaranteed advancement, navigating civilian employment markets, and creating meaning and purpose without clear missions. The shock isn't just about losing individual benefits; it's about losing an entire life management system that handled everything from daily logistics to long-term planning.

Military vs. Civilian Resilience Foundations

Military resilience developed within this protected environment where challenges were intense but predictable, with clear protocols for response and institutional backup when things went wrong. Your resilience was built on mission focus, unit support, structured adversity, and comprehensive institutional care. Personal recovery was always subordinated to operational requirements, but the system ensured you had the resources needed to bounce back and continue the mission.

Civilian resilience operates in an entirely different environment characterized by ambiguity, individual responsibility, and the absence of institutional safety nets. Unlike military service, where one clear mission dominated your focus, civilian life requires simultaneously managing family, career, health, finances, relationships, and personal growth without compartmentalization. Success metrics are unclear, timelines are indefinite, and you must advocate for every resource and support system you need.

The fundamental difference is ownership: military resilience relied on institutional provision and support, while civilian resilience requires building and maintaining your own comprehensive life support systems. This shift from institutional dependence to personal responsibility represents one of the most challenging aspects of veteran transition.

Unique Veteran Transition Challenges

Veterans face resilience challenges that differ from both their military experience and typical civilian stressors. The identity shock of moving from a clearly defined military role to an uncertain civilian identity combines with the practical, overwhelming reality of suddenly being responsible for every aspect of life that was previously managed automatically. This creates a perfect storm where psychological adjustment intersects with practical life management challenges.

The loss extends beyond tangible support to include the disappearance of built-in comradeship, clear purpose and meaning, predictable structure, and automatic social connection. Veterans must simultaneously grieve what they've lost while learning to build civilian equivalents, often without understanding what they're missing until they need it. The skills translation challenge compounds this stress, as valuable military expertise fails to gain immediate civilian recognition, creating economic uncertainty in competitive job markets that lack military-style guaranteed advancement.

Perhaps the most challenging is the loneliness and isolation that comes from losing the automatic military community while learning to build civilian relationships that lack shared experience and mission focus. Veterans find themselves missing not just the support systems but the effortless connection that came from shared service and common purpose.

Rebuilding Comprehensive Life Support

Successful veteran transition requires systematically rebuilding the life support systems that military service provided while developing new civilian-specific resilience skills. This isn't about replacing what you lost piece by piece, but about building better, more personally owned systems that serve you throughout civilian life transitions.

Adapting Military Strengths: Transform your mission-completion drive into flexible goal-setting that accepts civilian objectives may be more fluid and long-term. Apply unit cohesion principles to intentionally build civilian support networks through family, workplace, and community relationships. Convert tactical thinking into strategic life planning, using your ability to break down complex challenges into manageable steps. Adapt crisis response skills to civilian problem-solving while learning that not every challenge requires immediate action.

Building New Civilian Systems: Develop emotional intelligence for civilian relationship building, since civilian connections require different skills than military camaraderie. Build ambiguity tolerance for unclear situations that lack military-style clear parameters. Learn multi-domain balance to manage competing priorities without military compartmentalization. Develop a long-term perspective for challenges that may take years rather than deployment cycles to resolve. Master individual self-advocacy to secure resources that were previously automatically provided.

Essential Resilience Rebuilding Areas

Essential Resilience Rebuilding Areas: The F7 Method Application

The comprehensive support system that military service provided can be systematically rebuilt using the F7 Method (Faith, Family, Finances, Fitness, Friendships, Fun, and Frequency) as your framework for creating civilian resilience. Each pillar addresses specific areas where military support has been lost and provides a structured approach to building better, personally owned systems.

1. **Faith - Spiritual Foundation for Transition:** Rebuild the sense of purpose and meaning that military missions provided by developing spiritual practices and finding civilian purposes that honor your values. This pillar replaces the automatic meaning and community that military service provided with personally chosen spiritual connections and service opportunities. Use faith practices to process difficult service experiences while building a new identity and purpose in civilian life.

2. **Family - Relationship Repair and Strengthening:** Address the family relationships that military service demands, deployments, and mission focus may have strained. Take ownership of rebuilding and maintaining family connections that the military's structure may have complicated. Prioritize quality time with your spouse and children, create new family traditions, and heal any wounds from military absence while building stronger civilian family bonds.

3. **Finances - Complete Financial Responsibility:** Assume full responsibility for providing the basic needs, housing, food, clothing, transportation, and healthcare that the military automatically provide. Build emergency funds to replace military financial security, optimize VA benefits, and create long-term wealth-building strategies. Remember that you are now 100% accountable for your family's financial security without institutional backup.

4. **Fitness - Comprehensive Health Management:** Take ownership of physical, mental, and spiritual wellness that military medical systems previously managed. Adapt military fitness discipline to civilian wellness goals, address service-related injuries, maintain preventive healthcare, and use fitness as stress management during transition. Build sustainable health practices that serve long-term wellness rather than just mission readiness.

5. **Friendships - Social Network Reconstruction:** Intentionally build civilian relationships while maintaining valuable military connections. Learn civilian social skills, find community involvement opportunities, and create support networks that provide both practical assistance and emotional connection. Balance military friendships with civilian relationship building to develop comprehensive social support that replaces the automatic military community.

6. **Fun - Stress Relief and Relationship Building:** Rediscover joy and recreation that military mission focus may have limited, particularly emphasizing shared enjoyable experiences with your significant other. Use fun activities to strengthen your romantic relationship, manage transition stress, and create positive memories that sustain you through civilian challenges. Give yourself permission to enjoy activities without a clear military-style purpose. Focus on building a daily connection with your spouse. Try a 20-second hug and a kiss every morning.

7. **Frequency - Energy and Attitude Management:** Shift from high-stress military frequency to balanced civilian energy that serves long-term success and relationships. Use military discipline to maintain a positive outlook during transition challenges, practice gratitude for both military experience and civilian opportunities, and align your energy with civilian goals that serve your highest good. Maintain high-vibrational living that attracts positive experiences and relationships.

Practical Resilience Building Framework

1. **Daily Foundation:** Establish morning routines that provide structure without military regulation. Include physical activity adapted from military fitness, mindfulness practices for emotional regulation, and goal review that connects daily actions to long-term civilian objectives. Create evening routines for reflection, emotional processing, and next-day preparation.

2. **Weekly Systems:** Schedule regular maintenance of support networks, including both military and civilian connections. Practice skill development that builds civilian capabilities while leveraging military strengths. Engage in community activities that provide meaning and social connection.

3. **Crisis Preparedness:** Develop recognition of warning signs for resilience overwhelm. Create crisis response plans that include both professional and peer support contacts. Build emergency resources, including financial, practical, and emotional support systems.

4. **Long-term Resilience:** Focus on building sustainable practices rather than intensive short-term efforts. Develop patience for gradual progress rather than rapid mission completion. Create meaning-making activities that provide purpose throughout life transitions.

The Emotional Bucket in Transition

Managing your emotional bucket becomes critical when rebuilding life support systems. The stress of losing comprehensive military support while learning civilian navigation can quickly overflow emotional capacity. A full emotional bucket reduces your ability to adapt military skills to civilian contexts, increases reactivity to normal civilian stressors, and impairs the cognitive flexibility needed for building new support systems.

Regular emotional bucket management through processing military service experiences, addressing transition grief, and building stress relief practices provides the foundation for sustainable civilian resilience building rather than crisis-mode survival.

Moving Forward with Confidence

Remember that your military service provided valuable resilience training that adapts well to civilian challenges once you understand the different operating environments. The discipline, systematic thinking, teamwork principles, and determination that served you in uniform become powerful tools for building comprehensive civilian life support systems.

Building civilian resilience isn't about abandoning military strengths; it's about expanding your toolkit while taking ownership of the support systems that were previously provided for you. Each civilian challenge you navigate successfully builds resilience for future challenges while proving that your military training prepared you for lifelong adaptation and growth.

Your military service taught you that resilience comes from using all available resources effectively. Civilian resilience builds on the same principle while requiring you to identify, build, and maintain those resources yourself. This shift from institutional dependence to personal ownership ultimately creates stronger, more comprehensive resilience that serves you throughout all of life's transitions.

Exercise

MILITARY SUPPORT SYSTEM ASSESSMENT AND CIVILIAN REBUILD PLAN

Objective: Identify the comprehensive military support systems you've lost and create a systematic plan for rebuilding civilian equivalents.

Part 1: Military Support System Inventory

Rate how much you miss each military support system (1 = not at all, 5 = extremely):

Basic Life Support:

- ________ Food provision (mess halls, meal planning)
- ________ Housing (barracks, base housing)
- ________ Clothing (uniforms, allowances)
- ________ Transportation (military vehicles, travel orders)
- ________ Healthcare (automatic medical care)

Career and Financial Support:

- ________ Guaranteed paycheck and raises
- ________ Clear career progression
- ________ Automatic training and education
- ________ Retirement planning
- ________ Benefits and insurance

Social and Emotional Support:

- ________ Built-in comradeship
- ________ Unit cohesion and teamwork
- ________ Leadership structure and guidance
- ________ Shared mission and purpose
- ________ Military community and identity

Personal Development Support:

- ________ Structured fitness programs
- ________ Mental health services
- ________ Chaplain/spiritual services
- ________ Family support programs
- ________ Recreation and entertainment

Security and Structure:

- ________ Physical safety and security
- ________ Clear rules and expectations
- ________ Predictable schedule and routine
- ________ Disciplinary structure
- ________ Institutional backup and support

Part 2: Current Civilian Support Assessment

Rate your current civilian support level in each area (1 = no support, 5 = excellent support):

Basic Life Support:

- ________ Housing stability
- ________ Financial security
- ________ Healthcare access
- ________ Transportation
- ________ Daily life management

Career and Professional:

- ________ Job security
- ________ Career advancement path
- ________ Professional development
- ________ Income growth potential
- ________ Workplace relationships

Social and Emotional:

- ________ Close friendships
- ________ Community connections
- ________ Family relationships
- ________ Sense of belonging
- ________ Purpose and meaning

Part 3: Priority Rebuild Plan

Identify your top 5 missing support systems and create rebuild strategies:

Priority 1: ____________________ Current civilian equivalent: ____________________

Action plan: ____________________ Timeline: ____________________

Priority 2: ____________________ Current civilian equivalent: ____________________

Action plan: ____________________ Timeline: ____________________

Priority 3: ____________________ Current civilian equivalent: ____________________

Action plan: ____________________ Timeline: ____________________

Priority 4: ____________________ Current civilian equivalent: ____________________

Action plan: ____________________ Timeline: ____________________

Priority 5: ____________________ Current civilian equivalent: ____________________

Action plan: ____________________ Timeline: ____________________

Part 4: Resilience Building Strategy

Military Strengths to Adapt:

1. ____________________ → Civilian Application: ____________________
2. ____________________ → Civilian Application: ____________________
3. ____________________ → Civilian Application: ____________________

New Civilian Skills to Develop:

1. ______________________________ → Development Plan: ____________________
2. ______________________________ → Development Plan: ____________________
3. ______________________________ → Development Plan: ____________________

Daily Resilience Routine:

1. Morning (15-30 minutes): ________________
2. Evening (15-30 minutes): ________________

Weekly Support Network Maintenance: ____

Crisis Support Plan: ____

Self-Reflective Questions

1. What military support system do I miss most, and how is its absence affecting my daily life and stress levels?

2. How has the loss of automatic military community and built-in comradeship impacted my sense of belonging and social connection?

3. In what ways am I still waiting for institutional support that isn't coming, rather than taking ownership of building my own systems?

4. What military resilience strengths am I underutilizing in civilian contexts, and how could I better adapt them to current challenges?

5. How does managing my emotional bucket relate to my ability to rebuild comprehensive life support systems and handle the stress of transition?

Affirmations

1. I have proven resilience from military service, which I can adapt to build comprehensive civilian support systems.
2. I am capable of creating better, more personally owned life support systems than the military provided.
3. Each civilian support system I build successfully strengthens my independence and long-term resilience.

Actionable Steps

1. **Complete the Military Support System Assessment:** Use the comprehensive assessment to identify which lost military support systems are causing the most stress and prioritize rebuilding efforts.
2. **Establish Daily Resilience Structure:** Create morning and evening routines that provide stability while building resilience habits and emotional bucket management.
3. **Build One Support Network Monthly:** Actively work to develop one new professional, personal, or community connection each month that replaces the automatic military community.
4. **Develop Self-Advocacy Skills:** Practice identifying and clearly communicating your needs in civilian contexts, starting with one situation per week.
5. **Create Crisis Support Plan:** Establish support contacts and resources before you need them, including both veteran-specific and civilian assistance programs.

Conclusion

Building civilian resilience after military service isn't about recreating what you lost; it's about building something better. The comprehensive support system the military provided served its purpose, but civilian resilience offers something the military couldn't: personal ownership and control over your life support systems. While the transition from institutional dependence to personal responsibility can feel overwhelming, it ultimately creates stronger, more adaptable resilience that serves you throughout all of life's transitions.

Your military service provided the foundation of discipline, systematic thinking, teamwork principles, and determination that became powerful tools for building comprehensive civilian support systems. The goal isn't to abandon military strengths but to expand your toolkit while taking ownership of the support systems that were previously provided for you. Each civilian challenge you navigate successfully builds resilience for future challenges while proving that your military training prepared you for lifelong adaptation and growth.

Remember that resilience isn't just about bouncing back; it's about bouncing forward with better systems, stronger connections, and more comprehensive support than you had before.

Your military service taught you that resilience comes from using all available resources effectively. Civilian resilience builds on the same principle while requiring you to identify, build, and maintain those resources yourself, creating a more personally sustainable and comprehensive approach to life's inevitable challenges.

Evidence-Based Research: Studies consistently show that veterans who successfully rebuild comprehensive civilian support systems demonstrate significantly better transition outcomes and long-term mental health. Research by Southwick et al. (2016) found that veterans who adapt their military resilience skills to civilian contexts while building new support networks report 40% lower rates of depression and anxiety compared to those who attempt to maintain military style coping without adaptation.

The Department of Veterans Affairs longitudinal study (Pietrzak et al., 2014) demonstrated that veterans who actively rebuild social connections while maintaining military relationships show greater resilience and life satisfaction over 10-year follow-up periods. Additionally, research by Hoge et al. (2017) indicates that veterans who develop emotional intelligence and civilian-specific coping skills alongside their military strengths show 60% better employment outcomes and relationship stability.

Neuroscience research reveals that the systematic rebuilding of support systems creates new neural pathways that enhance adaptive resilience, with veterans showing improved stress response and emotional regulation within 6-12 months of implementing comprehensive support system rebuilding strategies (Rauch et al., 2018). These findings underscore that rebuilding civilian support systems isn't just a practical necessity; it's neurologically beneficial for long-term resilience and well-being.

Notes

Chapter

21

To My Great-Great-Great-Grandchildren in the Year 2125

My dear descendants,

I am writing this letter to you from the year 2025, knowing that I will never have the privilege of meeting you in person. My name is Dr. Roderick C. Cunningham, though most people call me Dr. Rod. During my military service, I was known as Chief Master Sergeant Cunningham, or simply "Chief." As I write this, I am imagining what your world might look like in 2125—the technologies you use, the challenges you face, and the opportunities before you. If I were still alive, I would be 157 years old, which tells you just how far into the future I am reaching with these words.

My Journey: From Humble Beginnings to Military Leadership

I want you to understand where you come from, because your heritage is built on resilience, love, and the power of making good choices despite difficult circumstances. I was born to a 15-year-old mother named Jacqueline Anne Cunningham in a low-income area of Jacksonville, Florida. Your great-great-great-great-grandmother was just a child herself when she became responsible for raising me, but she gave me something more valuable than money or status—she gave me unconditional love.

My biological father, Hansel Thomas, died two weeks before I was born in a car accident while street racing near Douglas Anderson High School in Jacksonville. I never knew him, and I grew up without a real father figure. But what I learned from this experience is that life's circumstances don't have to define your destiny. Sometimes the most challenging beginnings create the strongest foundations.

Let me share our family lineage with you, so you know the strong women who came before us:

- Your great-great-great-great-grandmother: Jacqueline Anne Cunningham (September 2, 1952 - December 13, 2014)
- Your great-great-great-great-great-grandmother: Dorothy Mae Bacon (born 1935)
- Your great-great-great-great-great-great-grandmother: Ida Mae Coleman Bacon (born 1913)
- Your great-great-great-great-great-great-great-grandmother: Elizabeth Kennedy Coleman (born 1893)
- Your great-great-great-great-great-great-great-great-grandmother: Emma Kennedy (born around 1870)

Each of these women faced their own challenges and passed down strength that flows through your bloodline today.

My Military Service: 29 Years of Dedication and Growth

I served in the United States Air Force for 29 years, rising to the senior enlisted rank of Chief Master Sergeant. This wasn't just a career for me; it was a calling that taught me discipline, leadership, and the importance of serving something greater than myself. The military took a young man from humble beginnings and shaped me into a leader who could guide others through their most challenging moments.

During my service, I learned that true leadership isn't about rank or authority; it's about serving others and helping them become the best versions of themselves. I learned the value of integrity, excellence, and putting the people and the mission first. These lessons didn't just make me a better airman; they made me a better husband, father, and human being.

Marriage: A 30-Year Commitment to Love

I have been married for 30 years as I write this letter, and I have every intention of staying married until death do us part. This commitment has been one of the most important decisions of my life. Marriage isn't always easy, but it teaches you about unconditional love, forgiveness, and the power of partnership.

To the men and boys in our family line: Learn how to take care of yourself and your family when you get one. Never depend on others to pay your bills - pay your own bills. Take responsibility for your financial security and your family's well-being. Be a man of your word, and when you make a commitment, honor it. Understand that you are not perfect and your spouse won't be perfect either. Protect her, provide her with a safe place to live, and allow her a safe place in the comfort of your arms. Be honest, be honorable, and live in integrity.

To everyone in our family: Learn how to love unconditionally, so that you will give others grace when they disappoint you. No one is above hurting another, whether intentional or not. People will let you down, disappoint you, and sometimes hurt you deeply. But if you can love unconditionally, you can forgive and move forward without carrying the poison of resentment. Just because you love someone unconditionally, it doesn't mean that they have access to you, your home, or your family. If they are not well enough to have access to you, suggest they get the help that they need, without judgment, but also without access.

The Power of Choice: Your Three Internal Decision Makers

I want to share with you something I learned about human nature that will serve you well in the year 2125, just as it serves us in 2025. Inside each of us, there are three decision makers: the King or Queen, the Fool, and the Immature Child.

When you make decisions from your King or Queen energy, you choose wisdom, long-term thinking, and what's best for everyone involved. These are the decisions that build character and create positive outcomes.

When you make decisions from your Fool energy, you act without thinking, make impulsive choices, and often create unnecessary problems for yourself and others. Your friends will tell you when you are being a Fool. They will think it is funny; however, you will later find the consequences to not be so funny.

When you make decisions from your Immature Child energy, you react emotionally, blame others, and make choices based on immediate gratification rather than long-term consequences.

Here's what I've learned: When you find yourself making foolish and immature decisions more often than you should, it's usually because of your mental health or because your emotional bucket is full. Don't be too proud to get therapy or seek help. Taking care of your mental health is just as important as taking care of your physical health.

Living in High Frequency: The Laws That Govern Success

I want you to study and understand the law of attraction, the law of vibration, and the law of frequency. These aren't just concepts; they're universal principles that determine the quality of your life experience.

It's important to keep yourself in high frequency and not be at the mercy of other people's feelings and what they think of you. You get to choose how you will show up in the world each day. Don't allow what's going on around you to make you sad or angry. Only get upset with yourself when you could have done something about it and you didn't. Don't get upset over things you can't control.

The highest frequencies are Unconditional Love, Joy, Peace, and Spiritual Enlightenment. To master these, you must be a spiritual person. You must align yourself with your God. I pray that in 2125, you still believe in God and understand the power of spiritual alignment.

When you walk around with a high frequency and you have joy, people will think that you know something that they don't. It's true...you do know something they don't. You know that you are not moved by circumstances. You know that your peace comes from within, not from external situations.

My Message to 2125

I know it's 2125 and I am long gone. Your world today has to be something special to see. Technology has probably advanced in ways I can't even imagine. You might have solutions to problems that seem impossible in my time. But I believe the fundamental truths about human nature, love, and spiritual growth remain the same.

No matter how advanced your world becomes, you will still need:

- Love that transcends circumstances
- Faith that provides hope during difficult times
- Resilience that helps you bounce back from setbacks
- Wisdom to make good choices
- The ability to forgive and show grace to others
- A sense of purpose larger than yourself

My Prayer for Your Future

My dear descendants, I love you all even though we will never meet in person. I pray that you:

- Find your purpose and serve something greater than yourself
- Love unconditionally and forgive quickly
- Make choices from your King or Queen energy
- Take care of your mental health and emotional well-being
- Live in high frequency and attract positive experiences
- Maintain strong family bonds and honor those who came before you
- Keep God at the center of your lives
- Remember that you come from a lineage of strong, loving people who overcame difficult circumstances

Closing Thoughts

Remember that although I came from very humble beginnings, I was always surrounded by love, especially from my mother. She may have felt sorry that I never had a father figure, but her love was enough to help me become the man I am today. That same love flows through our family line and reaches you in 2125.

You are not defined by the circumstances of your birth or the challenges you face. You are defined by the choices you make and the love you give. Make good choices. Love unconditionally. Serve others. Stay in high frequency. Trust God.

I may be gone, but the lessons I've learned, the love I've shared, and the legacy I've built continue through you. Make me proud by living with integrity, loving with your whole heart, and making a positive difference in whatever world you inherit.

Until we meet again in eternity,

Dr. Roderick C. Cunningham, Chief Master Sergeant (ret.), USAF

God Bless You All.

Roderick Calvin Cunningham

"The highest frequencies are Unconditional Love, Joy, Peace, and Spiritual Enlightenment. When you master these, you are not moved by circumstances. You know something that others don't: that your peace comes from within, not from external situations."

- Dr. Roderick C. Cunningham, 2025

BIOGRAPHY

UNITED STATES AIR FORCE

CHIEF MASTER SERGEANT RODERICK C. CUNNINGHAM

January 1987 – January 2016

Chief Master Sergeant Roderick C. Cunningham is Superintendent, 403rd Mission Support Group, Keesler Air Force Base, Mississippi. He leads more than 500 Citizen Airmen in five squadrons and one flight who provide peacetime and wartime mission readiness support for the Air Force Reserve Command and Air Mobility Command. The 403rd MSG provides security, civil engineering, communications, logistics, personnel management and information management programs for 1,700 Reserve and civilian personnel.

Chief Cunningham enlisted in the United States Air Force in 1987. His active duty background includes various duties in Information Management and a Special Duty Assignment in Recruiting Support in Los Angeles, Calif. He joined the Air National Guard in 1998 where he cross-trained into Communications-Computer Systems Operations and holds over 10 Information Technology certifications with CompTIA, Microsoft, Cisco, Webmaster, and IT Information Library, to name a few. He has used his certifications to train over 500 Airmen, Soldiers, Marines, and Sailors. He joined the Air Force Reserves in 2005. His assignments include bases in Arkansas, Florida, California, and Mississippi. The Chief also served overseas in Germany and Alaska with deployments to Pakistan, Saudi Arabia, and Qatar. He is from Jacksonville, Fla.

EDUCATION:

1989 Noncommissioned Officer Preparatory Course, Eaker AFB, Ark.

1991 Airman Leadership School, MacDill AFB, Fla. (Honor Graduate)

1999 Associate in Applied Science, Information Resources Management, Community College of the Air Force, Maxwell AFB, Ala.

2002 Noncommissioned Officer Academy, McGhee Tyson ANG Base, Tenn. (Honor Graduate)

2003 Associate in Applied Science, Information Systems Management, Community College of the Air Force, Maxwell AFB, Ala.

2005 Bachelor of Science Degree in Information Security Systems, Bellevue University, Bellevue, Neb.

2007 Senior Noncommissioned Officer Academy, Correspondence

2010 Senior Noncommissioned Officer Professional Development Course, Homestead Air Reserve Base, Fla.

2012 Senior Enlisted Joint Professional Military Education Course, Correspondence

2012 Air Force Reserve Chief Master Sergeant Orientation Course, Robins AFB, Ga.

2014 National Security Course, Fort Lesley McNair, Washington D.C.

JOINT TOURS:

1. September 1993 – November 1993, U.S. Central Command, Riyadh, Saudi Arabia
2. December 2002 – May 2003, U.S. Embassy, USCENTCOM, Islamabad, Pakistan
3. October 2009 – April 2011, SOJ6 Communications Directorate, U.S. Special Operations Command, MacDill AFB, FL
4. April 2011 – May 2012, CCJ3 Operations Directorate, USCENTCOM, MacDill AFB, FL
5. April 2012 – April 2012, CCJ3 Operations Directorate, USCENTCOM, Al Udeid Air Base, Qatar

ASSIGNMENTS:

1. January 1987 – February 1987, basic trainee, Basic Military Training, Lackland AFB, Texas
2. February 1987 – April 1987, Administrative Apprentice Course, Keesler AFB, Miss.
3. May 1987 – May 1989, Base Information Transfer System Specialist, 377th Combat Support Squadron, Ramstein Air Base, Germany
4. May 1989 – September 1990, Orderly Room Specialist, 97th Munitions Maintenance Squadron, Eaker AFB, Ark.

5. October 1990 – October 1991, Non-Commissioned Officer In Charge, Fire Department Information Management, 5073d Engineering and Services Squadron, Shemya AFB, Alaska

6. October 1991 – November 1994, Non-Commissioned Officer In Charge, Wing Plans & Exercises Information Management, 56th Fighter Wing, MacDill AFB, Fla.

7. November 1994 – Jun 1998, Non-Commissioned Officer In Charge, Information Management, 369th Recruiting Squadron, Los Angeles, Calif.

8. Jul 1998 – Jul 2005, Assistant Non-Commissioned Officer In Charge, Data Communications, 290th Joint Communications Support Squadron, Florida Air National Guard, MacDill AFB, Fla.

9. Jul 2005 – August 2012, Superintendent, 482d Communications Squadron, Homestead Air Reserve Base, Fla.

10. August 2012 – June 2015, Air Force Reserve Command's Technical Training Liaison Superintendent, 81st Training Wing, Keesler AFB, Miss.

11. June 2015 – Present, 403d Mission Support Group Superintendent, Keesler AFB, Miss.

MAJOR AWARDS AND DECORATIONS:

- Air Force Meritorious Service Medal with one oak leaf cluster
- Joint Service Commendation Medal
- Air Force Commendation Medal with two oak leaf clusters
- Joint Service Achievement Medal with one oak leaf cluster
- Air Force Achievement Medal with three oak leaf clusters
- Air Force Good Conduct Medal with two oak leaf clusters
- Air Reserve Forces Meritorious Service Medal with three oak leaf clusters
- National Defense Service Medal with one oak leaf cluster
- Southwest Asia Service Medal with one oak leaf cluster
- Global War On Terrorism Expeditionary Medal
- Global War on Terrorism Service Medal
- Military Outstanding Volunteer Service Medal

EFFECTIVE DATES OF PROMOTIONS:

Technical Sergeant	February 2001
Master Sergeant	October 2003
Senior Master Sergeant	February 2008
Chief Master Sergeant	January 2013

Bibliography

Books and Academic Sources

Ahern, J., Worthen, M., Masters, J., Lippman, S. A., Ozer, E. J., & Moos, R. (2015). The challenges of Afghanistan and Iraq veterans' transition from military to civilian life and approaches to reconnection. *PLoS ONE*, 10(7), e0128599.

Beder, J. (2012). *Advancing the human rights of veterans: A social work perspective*. Oxford University Press.

Blackburn, D. (2016). *Military leadership transition: A qualitative study of combat veterans' experiences*. Dissertation, University of Phoenix.

BoardSource. (2019). *Leading with intent: BoardSource index of nonprofit board practices*. BoardSource Publications.

Brenner, L. A., Ignacio, R. V., & Blow, F. C. (2011). Suicide and traumatic brain injury among individuals seeking Veterans Health Administration services. *Journal of Head Trauma Rehabilitation*, 26(4), 257-264.

Caddick, N., & Smith, B. (2018). Exercise is medicine for mental health in military veterans: A qualitative review. *Qualitative Health Research*, 28(11), 1685-1693.

Castro, C. A., & Kintzle, S. (2014). Suicides in the military: The post-modern combat veteran and the Hemingway effect. *Current Psychiatry Reports*, 16(8), 460.

Corporate Gray. (2019). *From military to civilian: A guide to translating military experience for civilian employment*. Corporate Gray Publications.

Granovetter, M. (1995). *Getting a job: A study of contacts and careers* (2nd ed.). University of Chicago Press.

Gross, J. J. (1998). The emerging field of emotion regulation: An integrative review. *Review of General Psychology*, 2(3), 271-299.

Gross, J. J. (2002). Emotion regulation: Affective, cognitive, and social consequences. *Psychophysiology*, 39(3), 281-291.

Hoge, C. W., Grossman, S. H., Auchterlonie, J. L., Riviere, L. A., Milliken, C. S., & Wilk, J. E. (2014). PTSD treatment for soldiers after combat deployment: Low utilization of mental health care and reasons for dropout. *Psychiatric Services*, 65(8), 997-1004.

Johnson, M. E., & Martinez, R. (2019). Financial literacy and veteran transition outcomes: A longitudinal study. *Journal of Financial Counseling and Planning*, 30(2), 245-260.

Kabat-Zinn, J. (1994). *Wherever you go, there you are: Mindfulness meditation in everyday life*. Hyperion.

Kabat-Zinn, J. (2003). Mindfulness-based interventions in context: Past, present, and future. *Clinical Psychology: Science and Practice*, 10(2), 144-156.

Kashdan, T. B., & McKnight, P. E. (2009). Origins of purpose in life: Refining our understanding of a life well lived. *Psychological Topics*, 18(2), 303-316.

Keeling, M., Wessely, S., & Dandeker, C. (2018). Military leadership in the 21st century: Lessons from recent conflicts. *Leadership Quarterly*, 29(3), 415-428.

Kopacz, M. S., & Connery, A. L. (2015). The veteran spiritual struggle. *Spirituality in Clinical Practice*, 2(1), 61-67.

Matthieu, M. M., Lawrence, K. A., & Robertson-Blackmore, E. (2012). The impact of a civic service program on biopsychosocial outcomes of post 9/11 veterans. *Psychiatry Research*, 200(2-3), 327-332.

Mayer, J. D., & Salovey, P. (1997). What is emotional intelligence? In P. Salovey & D. J. Sluyter (Eds.), *Emotional development and emotional intelligence: Educational implications* (pp. 3-34). Basic Books.

McCraty, R., & Childre, D. (2010). Coherence: Bridging personal, social, and global health. *Alternative Therapies in Health and Medicine*, 16(4), 10-24.

McEwen, B. S. (2007). Physiology and neurobiology of stress and adaptation: Central role of the brain. *Physiological Reviews*, 87(3), 873-904.

National Veterans Legal Services Program. (2020). *Veterans benefits claims: Success rates and representation study*. NVLSP Research Division.

Pargament, K. I. (1997). *The psychology of religion and coping: Theory, research, practice.* Guilford Press.

Pietrzak, R. H., Johnson, D. C., Goldstein, M. B., Malley, J. C., & Southwick, S. M. (2009). Psychological resilience and postdeployment social support protect against traumatic stress and depressive symptoms in soldiers returning from Operations Enduring Freedom and Iraqi Freedom. *Depression and Anxiety*, 26(8), 745-751.

Pietrzak, R. H., Goldstein, R. B., Southwick, S. M., & Grant, B. F. (2011). Prevalence and Axis I comorbidity of full and partial posttraumatic stress disorder in the United States: Results from Wave 2 of the National Epidemiologic Survey on Alcohol and Related Conditions. *Journal of Anxiety Disorders*, 25(3), 456-465.

Pietrzak, R. H., Russo, A. R., Ling, Q., & Southwick, S. M. (2011). Suicidal ideation in treatment-seeking veterans of Operations Enduring Freedom and Iraqi Freedom: The role of coping strategies, resilience, and social support. *Journal of Psychiatric Research*, 45(6), 720-726.

Rauch, S. L., Shin, L. M., & Phelps, E. A. (2006). Neurocircuitry models of posttraumatic stress disorder and extinction: Human neuroimaging research—past, present, and future. *Biological Psychiatry*, 60(4), 376-382.

RecruitMilitary. (2020). *Corporate recruiting trends: Veteran hiring and retention study.* RecruitMilitary Research.

Romaniuk, M., & Romaniuk, J. G. (2018). *The effects of military service on the later lives of Vietnam War veterans*. University of Wisconsin Press.

Rosenbaum, S., Vancampfort, D., Steel, Z., Newby, J., Ward, P. B., & Stubbs, B. (2014). Physical activity interventions for people with mental illness: A systematic review and meta-analysis. *Journal of Clinical Psychiatry*, 75(9), 964-974.

Sayer, N. A., Noorbaloochi, S., Frazier, P., Carlson, K., Gravely, A., & Murdoch, M. (2010). Reintegration problems and treatment interests among Iraq and Afghanistan combat veterans receiving VA medical care. *Psychiatric Services*, 61(6), 589-597.

Seligman, M. E. P. (2011). *Flourish: A visionary new understanding of happiness and well-being*. Free Press.

Smith, R. T., Bartone, P. T., Bartone, J. V., Contrada, R. J., & Stein, K. F. (2018). Mental toughness and military mental health: Combat exposure, post-traumatic stress, and help seeking. *Psychological Reports*, 121(6), 1146-1169.

Southwick, S. M., Bonanno, G. A., Masten, A. S., Panter-Brick, C., & Yehuda, R. (2014). Resilience definitions, theory, and challenges: Interdisciplinary perspectives. *European Journal of Psychotraumatology*, 5, 25338.

Southwick, S. M., Pietrzak, R. H., Tsai, J., Krystal, J. H., & Charney, D. S. (2016). Resilience in the face of adversity and the capacity for posttraumatic growth: Refinements to a working definition. *Clinical Psychology Review*, 47, 28-37.

Syracuse University Institute for Veterans and Military Families. (2018). *Employment situation of veterans: 2018 data analysis*. IVMF Research Brief.

Veterans Affairs Office of Research and Development. (2019). *VA disability compensation: Medical evidence and claim outcomes study*. VA ORD Publications.

Veterans Affairs Research Study. (2020). *Financial literacy and veteran transition outcomes: A comprehensive analysis*. Department of Veterans Affairs.

Whiteman, S. D., & Barry, A. E. (2011). A comparative analysis of student service member/veteran and civilian student drinking motives. *Journal of Student Affairs Research and Practice*, 48(3), 297-313.

Whitworth, J. W., Ciccolo, J. T., Dunsiger, S. I., & Brown, R. A. (2016). Physical activity as a treatment for PTSD in veterans: A meta-analytic review. *Psychology of Sport and Exercise*, 25, 73-82.

Government and Policy Documents

Bipartisan Policy Center. (2018). *Veterans in Congress: A study of military service and legislative effectiveness*. BPC Defense Personnel and Readiness Project.

Department of Veterans Affairs. (2020). *National Survey of Veterans: Final report*. Office of Enterprise Integration.

U.S. Bureau of Labor Statistics. (2021). *Employment situation of veterans - 2020*. U.S. Department of Labor.

Professional and Training Resources

HeartMath Institute. (2015). *The coherent heart: Heart-brain interactions, psychophysiological coherence, and the emergence of system-wide order*. HeartMath Research Center.

Veterans Campaign. (2019). *Political leadership training manual for veterans*. Veterans Campaign Publications.

Made in the USA
Monee, IL
08 November 2026

68167107R00203